Anonymous

The Natural History of Birds

Anonymous

The Natural History of Birds

ISBN/EAN: 9783337816520

Printed in Europe, USA, Canada, Australia, Japan

Cover: Foto ©berggeist007 / pixelio.de

More available books at **www.hansebooks.com**

THE

𝔑atural 𝔥istory

OF

BIRDS;

CONTAINING

A VARIETY OF FACTS,

SELECTED FROM SEVERAL WRITERS,

ILLUSTRATED WITH

Upwards of One Hundred Copper-plates.

IN THREE VOLUMES.

VOL. III.

LONDON:

PRINTED FOR J. JOHNSON, NO. 72, ST. PAUL'

CHURCH-YARD,

THE

NATURAL HISTORY

OF

BIRDS.

ORDER 6th.   PASSERES.

The beak is conic, and sharp at the point.
The nostrils are oval, wide, and ...

THIS Order may be separated into two
divisions.

First, the Columba, or Columbine,
which includes Pigeons.

Secondly, the Passerine, which includes all the
other Genera that are placed under the Order of
the Passeres.

Strictly speaking the Genus, or rather the Division Columba, forms an order of itself, which
has been called the Columbine; for as is justly
observed in a note in Linnæus, Pigeons cannot
properly be classed either with the Gallinæ, or

the

the Pafferes, becaufe they pair, they bill, they fit alternately, they lay fewer egg, they feed their young from their own beaks with what they have fwallowed, and they build their nefts on high.

However as Linnæus, whofe arrangement is followed, has placed them with the Pafferes, they will be confidered here as belonging to the firft divifion of that order.

Firft Divifion of PASSERES.

Genus 63.   COLUMBA.

The bill is foft, nearly ftrait, defcending a little towards the point.

The noftrils are oblong, half covered with a foft tumid or fwelling membrane.

The tongue is intire.

THE genus Columba, or Pigeon, admits of two fubdivifions. The firft with the feathers in the tail nearly of an equal length.

It has been obferved that Pigeons are not our fervants, like dogs and horfes, nor are they our prifoners like fome kinds of poultry, but that they rather feem to be our voluntary companions, and occafional guefts, fo long as they are provided

with

with the conveniences they like.—S·me p···t
the dufty holes in decayed wall to ···· ···· t
boxes we can furnifh them with. S····· in
the crevices of rocks, or hollows of t··· s, w··d
others never abandon thofe places that are ap-
pointed for them.

All animals in a domeftic ftate become varioufly
marked, and from being better fed will be increafed in fize; thefe differences by attention may
be rendered conftant, and varieties without nt
may, by culture and attention, be procur·d from
one fpecies.   This is inftanced in white Turkeys,
white Ducks, Ducks and Fowls that are crefted,
but in no birds does it prevail in a greater extent
than amongft Pigeons, which, being very prolific,
that is breeding very faft, would naturally become the moft varied.   Some perhaps have been
felected on account of their fuperior fize, and from
thefe a larger race has been produced; others on
account of particular marks have been matched
together, and thefe marks have become perma-
nent or lafting.   Others again, on account of fome
of their feathers growing in a particular form, or
taking a particular direction.   Moft of thefe, in
the firft inftance, were only accidental varieties,
the confequence of their becoming domeftic.

A 3   The

The effects of cultivation are seen more or less in every animal under our care. Canary birds are now bred with crests, some with yellow feathers, some mottled, and by combining these varieties other new varieties will probably take place.

An account then of every different race of Pigeons, is not so properly the natural history of the bird, as an account of the art and industry of man, applied to this particular object.

The Stock Dove is so called because it is the stock or stem from which probably most of the others sprang; for however numerous the species of Pigeons may appear, it is very possible that a few only exist in nature, and that the rest are varieties, the natural effect of domestication at first, and particular attention afterwards.

The Stock Dove, which is the Pigeon in its wild state, breeds but once or twice in the year; they migrate into England in great numbers from colder climates at the approach of winter, some indeed stay with us the whole year, breeding in woods, on trees, in the holes of trees that are decayed, or in the crevices of rocks, whence it has sometimes been called the Rock Pigeon, and the Wood Pigeon.

Whilst

Whilst the beech woods covered large tract of
land, they haunted them in myriads, their flocks
extending in strings of a mile in length as they
went out to feed.

Their plumage is generally of a bluish cast, the
neck glossed with green, the back is white, and
there is a black band on each wing, and at the
end of the tail.

The first stage of domestication is the Dove-
house Pigeon; this is the intermediate state between the wild and the tame Pigeon. The Dove-
house Pigeons breed three times a year, if they
are pleased with their situation. As they dislike
interruption, a dove-house situated close to our
habitations seldom succeeds so well as one at the
distance of four or five hundred yards, and exposed to the morning sun.

They frequently desert their houses when their
situation is low, and fly to take possession of others
whose exposure is more pleasant; but if they
should continue in those which have not the be-
nefit of the morning sun, they would not increase
so fast.

A Pigeon-house in a fine situation has produced four hundred pairs in a season, whilst another, less favourably placed, has produced only
one hundred pairs.

A 2

When Pigeon-houses are at some distance, it is necessary to guard them from birds of prey, such as Hawks and Owls, and even from Magpies.

The tame Pigeon is in a still higher state of domestication, its colours and varieties are infinite.

These Pigeons will breed almost every month, provided there be not too many in the same Dove-house; but then each pair must have three or four boxes, and so contrived that they may not see one another whilst they sit; if the number be too great, they will be frequently fighting and breaking the eggs.

They begin to lay at nine months old, and sit from seventeen to twenty days; and as they breed so frequently, it has been calculated that in four years one thousand eight hundred might be produced from one pair.

They lay each time two eggs, scarcely making any nest; the cock and hen sit by turns, one calling to the other by a gentle cooing, whenever it wishes to leave the nest in search of food. The cock watches near his mate attentively, whilst she is sitting, and takes her place for two or three hours at a time, sometimes twice in a day and night.

Their

Their fondness for their e_gs i _ _ _ _, that
they will suffer severe hardships, rather than forsake them. A hen Pigeon, whose box was un-
fortunately placed so near a window during some
very cold weather, that her feet were frozen and
fell off, still continued sitting, until she had hat-
ed her young.

The young Pigeons are generally a cock and a
hen; their parents feed them by first swallowing
the grain, and after it is a little softened in their
first stomach, they threw it back thrc__ their
beaks into the beaks of their young.

After Pigeons are once paired, they are very
constant: nothing can be more _____ and assi-
duous than the attention of the mare to his partner, he seems to sollicit her regard by the most
interesting manners, he walks frequently round
her, displaying himself in a variety of attitudes,
and calling to her in a note the most tender and
plaintive. Of all birds they appear the most happy
and affectionate, cooing to, and billing one another
in a manner that seems to discover a tenderness,
and even a delicacy of attachment.

The Greeks had improved the breed of Pigeons,
and kept them in Dove-houses. Pliny, among
the Romans, speaks of a large breed of Pigeons
in Campania, and of people who, being very cu-

rious in Pigeons, bought them at extravagan
prices; he fays that they built turrets for them at
the top, of their houfes, and kept their pedigree.

It was probably from the Romans that we
might firft have learnt to keep them in Dove-
houfes. In Egypt they are very frequent, and are
confidered by the hufbandman as a valuable part
of his eftate.

Pigeons are very fond of falt, and have fre-
quently been the means of pointing out falt fprings.
In order to attach them to any particular Dove-
houfe, a compofition is fometimes made of falt,
caraway-feeds, loam, and rubbifh; and this is
occafionally ufed to intice other Pigeons.

They are very delicate food.

Amongft the many varieties the following are
fome of the moft remarkable.

The Roman, or Spanifh Pigeon, it has a cere
of a white mealy appearance, and is double the
fize of a common Pigeon.

The Rough-footed Pigeon, whofe legs are co-
vered with feather to the toes.

The Powter Pigeon, he can inflate or blow up
his breaft to a monftrous fize, as large as the reft
of his body. Linnæus fays, that this variety, or
fpecies is found in Arabia Felix. The pureft
breed

breed of thefe have the ten quill feathers white as far as the middle of the wing.

The varieties of this Pigeon are very nume rous. As their breaft is almoft always filled out with wind, they are obliged to draw back their heads; this prevents their feeing before them, and they are often furprifed by birds of prey.

The Jacobine. The feathers round the back part of the head are raifed, and turn in a contrary direction; the beak is very fhort. Of it there are many varieties; in general they breed pretty well, and they are beautiful birds: the Puff and Capuchin Pigeons feem nearly related to the Jacobine.

The Laced Pigeon is white, the legs red, the feathers are loofe in their webs, and curled or frizzled.—The Frillback is nearly related to it, the tip of each feather being bent upwards.

The Turbit Pigeon.—The feathers of the breaft on each fide turn back, the beak is fhort, and the head fmooth; this is a very beautiful bird, but does not eafily pair with others, and is frequently taken by birds of prey.—The Owl Pigeon is like this, except that it is only of one colour, and the Turbit is of two colours.

The

The Fantail Pigeon.—The tail is raised, very wide, and furnished with many feathers. This Pigeon can raise and spread its tail like the Turkey Cock. Some of them have thirty feathers in their tails, whilst many other Pigeons have but twelve. When their tails are spread, they are not only raised but thrown forwards towards the head, and their heads are drawn back towards the tail; all this time there is a quivering in the neck, as though it were convulsed by the seeming exertion. These Pigeons do not fly so well as others, their large tails appear to be an incumbrance, and often occasions them to be blown out of their right direction by the wind. Some of these are quite white, and some white except the head and tail, which are black.

The Tumbler is rather smaller than the common Pigeon, and frequently turns over and over as it flies, until it almost reaches the ground, as though it were seized with a giddiness. It rises very high, its flight is rapid, and all its movements quick and irregular. It is a small Pigeon, and has sometimes been called the Pantomime Pigeon from its tumbling.

The Almond Tumblers are the most valued: eighty guineas have been given for one of them.

There

There is a Pigeon nearly allied to the Turbler, which, as it turns round, strikes the air with that force with its wings, as sometimes to break several of its quill feathers.

The Helmet Pigeon has the head, the tail, and the first quill feathers of the same colour; the rest of the feathers are of a different colour.

The Persian Pigeon has a large warty excrescence above the beak, and a red circle extending round the eyes; it is a large Pigeon, its legs are short, and it seldom goes far from the Dovehouse.

The Carrier has a remarkably large white carrunculated or warty cere, and a naked circle round the eyes.

These Pigeons have often been employed to carry letters to a considerable distance in a short time.

In the Pigeon-houses of Grand Cairo, in Egypt, they separate some of the males from the females, and send them to those places to which they wish to receive news; the letter is tied under the wing of the Pigeon, who has been previously well fed, lest he should stay by the way; he flies strait to the Pigeon-house where his mate is kept, and in one day paffes over a fpace that a man would be fix days in travelling.

At Aleppo they have Pigeons which will bring letters in lefs than fix days from Alexandria, although it be four hundred miles diftant. It is faid that a Pigeon went from Babylon to Aleppo, which is thirty days journey, in forty-eight hours.

At the fiege of Modena, Pliny mentions, that Hirtius without, and Brutus within the walls, kept up a conftant correfpondence by Pigeons; and by that means baffled every attempt of Antony, (who was befieging the place,) to intercept their letters.

In England they are often let loofe at Tyburn to inform the diftant friends of the criminal of his difgraceful death.

They do not always ufe on thefe occafions the Carrier, which is a valuable Pigeon, as many others will anfwer the purpofe.

. The Horfeman is a breed between the Carrier and Powter, and often employed to convey letters.

A Dragoon Pigeon, which is bred by a Carrier and Horfeman, flew from St. Edmund's Bury to Bifhopfgate-ftreet, which is feventy-two miles, in two hours and a half.

The

The Spot Pigeon has a spot on the forehead of the same colour as its tail; the middle tail feathers are white.

The Norway Pigeon has a red head, and long feathers on its legs; it is as large as a fowl, and white as snow.

The Barbary Pigeon has a short bill, a broad circle of mealy and warty red flesh round its eyes; the plumage is bluish, and it has two blackish spots on each wing.

There are probably many other varieties already known to Pigeon-fanciers, and many new ones, perhaps, forming every day.

The Partridge Pigeon has a naked red skin round the eyes, the body is reddish above, yellow beneath.

It has a white mark on each side under the eye, on each side the throat, and at the setting on of each wing.

It

It inhabits Jamaica, feeds on myrtle-berries, builds in trees that have low boughs, and lines its nest with hair and cotton: at firft fight it feems much like a Partridge.

The White Crowned Pigeon.—The crown of the head, and the circles round its eyes are white, the feathers of the body bluifh, the wing and tail feathers are dufky, the beak red at the bafe.

The White Crowned Pigeons inhabit North America, the Bahama iflands, and Jamaica; the inhabitants take great numbers, they feed on the feeds of the mangrove and wild coffee, and are bitter or fweet, according to the food they live upon.

The White-winged Pigeon has a naked blue fkin round the eyes, the tail feathers are afh-colour with white tips, the two middle tail feathers dufky. It inhabits Afia, and flirts its tail like a Water Wagtail.

The Triangular Spotted Pigeon has a naked red fkin roun' each eye, triangular white fpots on the wings, tail feathers black at the tips; the bill is dufky. It inhabits the fouthern parts of Africa, and is common about the Cape of Good Hope.

The

The Great Crowned Pigeon has a black circle round each eye, an upright creft, the body bluifh, the fhoulders ferrugineous or iron colour.

Six of the covert feathers of the wings are black, but iron-coloured at the points. The creft on its head is large, upright, comprefied, and compofed of diftinct feathers; the bill and legs are dufky, and the bird is nearly the fize of a Peacock.

It inhabits the Molucca iflands and New Guinea, and has been brought alive to England. In the Eaft Indies they are fometimes kept tame in their poultry-yards.

The male approaches his partner with the fame geftures as the common Pigeon, its note is cooing and plaintive, but loud. The crew of Mr. Bougainville, (who made a voyage round the world,) were much alarmed at their noife, imagining before they difcovered what it proceeded from, that it was the cries of human beings. They build their nefts on trees.

The Ring Dove.—The tail is afh-coloured above, and blackifh at the tips; the greater quill feathers are dufky, and all except the outermoft have the exterior edges whitifh; on each fide of the neck is a white crefcent; the laft re cove ed with feathers, almost to the toes. They inhabit Europe and Asia, fly in flocks, and are very hurtful to corn-fields.

The Ring Doves are a large species of the Pigeon; they come into England and France in the spring, and go away in autumn. Their young are seen in great plenty in August. Some continue in France all winter; they perch upon trees, and make a large, but a slight and flat nest among the branches. They lay two, and sometimes three eggs, and breed twice a year, in April and at Midsummer. Mr. Buffon says, that they sit but fourteen days, and that in fourteen days after the young are hatched, they are capable of taking care of themselves. They coo louder than the common Pigeon in the breeding season when the weather is fine, but if it be rainy they are quite silent.

They feed upon berries, acorns, wild strawberries, and grain of all kind, and frequently do great mischief among the corn.

They are very delicate food, and being much sought after, are rather scarce. In some parts of England they are called Queists, and their note is most pleasingly wild and plaintive.

The White-bellied Pigeon.—The tail is blue, terminated with a band of white. It is found in Jamaica,

Jamaica, and is probably a bird of paffage; it feeds on berries, perches on trees, and makes a mournful noife, at the fame time that it is loud and difagreeable.

The Turtle.—The tail feathers are white at the points, the back is greyifh, the breaft is a pale reddifh colour; on each fide the neck is a fpot of black feathers, with white tips.

Linnæus confiders this bird as an inhabitant of India; it is not uncommon in England and in France. Turtles are very fhy birds, retiring to breed in thick gloomy woods, generally amongft oak trees; during that feafon they are found in Buckinghamfhire, they afterwards migrate into Gloucefterfhire, Shropfhire, and into the weft of England; but in autumn leave the kingdom. Indeed they arrive later, and depart fooner, than any other fpecies of Pigeon. In the fummer they frequent the pea-fields in Kent in flocks, as foon as the peas begin to ripen, and deftroy great numbers. The female lays two eggs, and is fuppofed to breed but once in the feafon.

The Collared Turtle.—The body is of a light clay-colour; the hind part of the neck is marked with a collar of black feathers. Linnæus fays that
this

this bird is found in India; it is however a bird of
paſſage in moſt European countries.

The Collared Turtles migrate in flocks; they
continue in France four or five months, and bree
there, building their neſts in the moſt ſilent an
gloomy woods. Though naturally they are wilde
than Pigeons, yet, like them, they may be tamed
and brought up in Dove-houſes. They may be
paired with the common Pigeon, and a mixt
breed will be produced, but it is not known whether theſe will produce others.

In ſome countries not frequented by men,
Turtles are ſo tame, that they may be beat down
by hundreds with a ſtick. Though their manners be gentle, careſſing, and affectionate, they
are not ſo conſtant to each other as the common
Pigeon. They eat and drink like the reſt of the
Pigeons, not raiſing their heads until they have
ſwallowed all the water they want at that time
Their cooing is ſweetly melancholy and plaintive

The Ground Turtle.—The tail and quill feathers are duſky, the body purpliſh, the beak and
feet yellowiſh. This ſpecies is about the ſize o
a Lark, and inhabits the warmer parts of America; feeds on the ground like Partridges; when
diſturbed they take a ſhort flight, alighting again
ſoon

roon upon the ground; they eat grain of various kinds, and are frequently taken in traps, being reesteemed delicate food.

The Pafferine Turtle is still lefs, its plumage is dusky, it has five fpots on each wing, like polifhed fteel; the outer tail feathers are white at the tips, the breaft is a whitifh red, and the wings reddifh beneath. It inhabits the ifland of St. Domingo, and may be a variety of the Ground Turtle.

The Malacca Turtle is but a very little bigger than a Houfe Sparrow; the bill is black, tinged with yellow at the tip and bafe; fides of the neck white, croffed with many lines of black; outer tail feathers brown two thirds of their length, from thence to the end white; legs a yellow orange-colour.

It inhabits Malacca, is a beautiful fpecies, and delicate food. Some of them have been carried to the Ifle of France, and increafe there very faft.

The Pompadour Pigeon has a bluifh bill, its cheeks and chin are of a pale yellow; the back, breaft and ftomach, a pale green; the wing coverts a beautiful rich red, the quills black edged with yellow; the tail of a light green and long. It inhabits Ceylon, and is always feen flying or

perched

perched on trees; its food is berries. The natives catch these Pigeons with bird-lime, the Europeans shoot them. Their flesh is esteemed a delicacy.

The Purple Crowned Pigeon.—Of this several varieties are found in the islands in the South Seas or Pacific Ocean. In Otaheite the crown of this Pigeon is a faint purple; at Ulietea deeper; at Tonga Taboo of a very deep vivid purple, surrounded by yellow. They feed on the fruit of the banana, and are easily tamed.

Mr. Bougainville mentions a beautiful green Turtle, and Pigeons with a green and gold plumage, perhaps of this species.

The White Nutmeg Pigeon.—The plumage is wholly white, except the quills, and one third of the tail near the end, which are black; the legs are light grey.

It inhabits New Guinea, feeds on nutmegs, the outer skin of them is perhaps the only part that serves for nourishment, as the nut itself passes through the bird uninjured for the purposes of vegetation; and by this means, the nutmeg-trees are propagated, and spread through all the islands which they frequent.

4·   There

There are other fpecies, or varieties, that are called Nutmeg Pigeons for the fame reafon. Nor is this the only fpice that is propagated in this manner. The cinnamon groves are diffufed in Ceylon by the fame means, by Pigeons called Cinnamon Eater; and there are other plants the berries of which they eat, voiding the ftones on trees and on the ground, where they grow.

The Hackled Pigeon's bill and irides are crimfon, the feathers of the head, neck, and breaft, are long, narrow, and pointed, and produce a very fingular effect, refembling a polifhed fubftance; it has a naked fkin round each eye, of a deep red colour; the back, wings, and ftomach, are of a deep blue; the tail a deep crimfon, and the legs black.

They are found in the Ifle of France, and their flefh is fuppofed to be poifonous.

The Second Divifion, or PIGEONS, the feathers in whofe tails are of unequal length.

The Paffenger Pigeon.—The tail is wedgefhaped, there is a naked crimfon circle round each eye, and the breaft of a reddifh brown.

Thefe Pigeons inhabit North America; their migrations depend very much on the feafon, for

it

it is the want of food that compels them to fly from one province to another. If the winter be mild, few are feen in the fouthern parts of that continent. They feed on acorns, the berries of various trees, and feveral kinds of maft. Wherever they flop they devour every thing of that kind, for they fly in myriads.

When they have eaten all the mafts, or berries, that are fallen upon the ground, they fly ftrait upwards amongft the branches of the trees, in columns, by turns, and they beat down the acorns and maft with their wings. In Canada they frequently do great mifchief in devouring the corn, before they begin their flight into more fouthern countries. They abound in the country near Hudfon's Bay, where they breed in the woods ; in fome parts, it is faid, they are in fuch numbers, that they join neft to neft, and that thefe nefts reach from tree to tree for feveral miles in length. When their food is exhaufted in thofe cold climates, they collect in immenfe numbers, and it is afferted that they fly by millions in a flock, intercepting the light of the fun, fo as to creafion a degree of darknefs in thofe places over which they pafs. Thefe flocks are faid to be two miles long, and a quarter of a mile in breadth.—According to fome writers, they are four or five

miles

miles in length, succeeding one another almost continually for three days together.

They alight in such numbers upon trees, as frequently to break off by their weight limbs of the largest oaks, and the ground under the tree where they have roosted, is covered to a considerable depth with their manure.

In North America the Indians watch their roosting places, go there in the night with long poles, and kill them by thousands. Formerly in almost every little town in the interior parts of Carolina, the inhabitants would provide a hundred gallons of the fat or oil of these Pigeons, which they eat with maize instead of butter.

A gentleman thought of a method of taking these Pigeons in great numbers, and with little trouble. He placed a vessel with burning sulphur under the trees, where they roosted; the fumes suffocated them, and brought them down in showers.

In Philadelphia the inhabitants often shoot them from the tops of their houses.

It is very singular that some days, when almost all the inhabitants of a town go out to kill Pigeons, and destroy great numbers, that not one hen Pigeon shall be found amongst them; and on other days not one cock.

PART VI.　　B　　The

The Carolina Pigeon.—The tail is wedge-shaped, and very long; it has a blue circle round each eye; the feathers on the under part of the body are a reddish brown.

It inhabits Carolina all the year, and feeds on berries and feeds. It is a great devourer of peas, for which reason, and because its flesh is delicate food, great numbers are taken in traps.

There are some other species of Pigeons of this as well as of the former division, but nothing very interesting is known respecting their habits; they are only distinguished by natural historians by the difference of their plumage, &c.

ORDER 6th. PASSERES.

Second Division, the PASSERINE.

The beak is conic, and sharp at the point.
The legs formed for hopping, and slender, the toes are divided.

GENUS 67. ALAUDA.

THE LARK.

The beak is cylindrical, and slender; strait except towards the point, where it is slightly bent; the mandibles equal, gaping outwards at the base.

The tongue is divided at the end.
The hind claw nearly strait, and longer than the toe.

THE SKY-LARK.

The two outer tail feathers are white on the outer web their length, the inner webs of the intermediate tail feathers, are of a rust colour.

SKY-LARKS inhabit all parts of Europe; and of Siberia, as far as Kamtschatka; removing from cold into milder climate, at the approach of winter. In the neighbourhood of Dunstable, four thousand dozens have been taken between the 14th of September, and the 25th of

B 2 February,

February, for the London market. The Lark in the country near Leipfic, are as famous in Germany, as thofe of Dunftable are in England, and fuch numbers have been brought to Leipfic fair, that the excife or tax, that has been paid upon them, has amounted to 6000 dollars.

Nor is it only in Europe, and Afia, that Skylarks are found. They likewife inhabit Africa but hitherto we have not heard that any of this fpecies have been feen in America. In the fummer, they prefer high, and dry fituations; in the winter they defcend into the plains, or level felds; it is then that they are fatteft, for they are almoft continually feeding; but in fummer they fly, and fing fo much, and are fo much engaged with the care of their young, that they are always lean.

The male bird is crefted, at leaft it has the power of raifing fome of the feathers which grow upon its head; and it can learn the fong of almoft any other bird, even after its own is in fome meafure formed; it may be taught tunes by a bird organ, and performs them better than either a Canary-bird, or Bullfinch: it is fo ftrongly difpofed to imitate, that in order to have its own or any acquired fong unmixed, it ought to hear no other.

If

In a ftate of liberty the Sky-lark begins his long, the firft fine weather in the fpring, generally early in the morning, and repeats it in the afternoon: he rifes high in the air, fometimes remaining as it were fufpended, then rifing again by jerks, he warbles the moft mufical, and animated notes; finging ftill louder as he rifes higher, fo that his fong may be heard when he has foared out of fight; and lowering it as he defcends, until it dies quite away.

He never fings on the ground, though he lives moftly there, for the Sky-lark cannot perch upon a tree, the form of his nail behind, which is nearly ftrait, and very long, preventing him from clafping the branch.

Though Sky-larks generally rife in the air by jerks, yet they frequently defcend obliquely; and fometimes, (efpecially when they are in fear of a bird of prey,) they fall inftantly like a ftore.

Like other birds that mount very high, Larks have been carried away to fea by the wind, and accordingly we read many inftances of Larks, and other birds, alighting on the mafts of fhips, at a great diftance from land. By extreme cold weather, they are forced to the neighbourhood of fprings that do not freeze, in large flocks; in -.

... in winter they are always collected together
in ... ...

In the cold weather of January 1776, in the
... ... of Pont de Beauvoisin in France,
they appeared in such numbers, that a man with
his kids could ... ... two mules; they were very
... ... almost starved for want of food

The female makes her nest on the ground,
and lines it with grass, and small dry roots, and
fibres: she conceals it with so much care, that it
is found with great difficulty. She sits only ele-
ven days, and in fifteen more, the young ones
leave the nest, and follow her.

As the nest is made, and the young are hatched,
and capable of providing for themselves, in so
short a time, it is probable that they breed more
than once in a season. In warm climates they are
supposed to hatch three times; in England perhaps
twice; in Sweden, which is colder, only once.

The mother does not often warm her young ones
under her wings, they lie scattered at a little distance from one another, and she often flutters
above them.

Buffon mentions an instance of a young lark
which he had in confinement, that conceived
all the affection of a parent for some nestlings
that were brought to her, and though she was

but

but a little elder than their young ones, yet she
nursed them with a mother's attention, and anxiety.

After a separation, she flew to them the moment she was at liberty; and though the opportunities of escaping were frequent, she was
withheld by her attachment. Her anxiety was so
great, that at length she died of fatigue; the
little ones did not long outlive their affectionate
nurse, whose care had been so useful to them.

Young Larks feed on worms, ants eggs, caterpillars, and grasshoppers; for which reasons they
are much esteemed in the countries infested by
these insects.

They are taken in many different ways; when
the weather is gloomy, they are caught in a net
which is stretched on two poles eighteen feet
long, and carried by two men. Sometimes they
are caught by a long string, part of which is covered with bird-lime, and let drop upon the bird
by two men, or boys, who walk along the field at
as great a distance from each other, as the length
of the line will permit, each holding one end in
his hand.

Sometimes they are enticed by call birds of
their own species, (for we cannot deceive them
by imitating their notes,) or by a looking glass
fixed in wood, which by means of a string is made

to

to turn fwiftly round; perhaps the light it re-
flects may raife their curiofity.

When they are caught in order to be confined,
as finging birds, it is fometimes neceffary at firft
to tie their wings, until they are a little ufed
to the cage; or to ftretch a piece of filk juft
beneath the upper part of the cage, becaufe they
are apt to fly againft it violently, and to hurt
themfelves.

Their cages fhould be very large, and perches
are unneceffary, but a turf fhould always be
placed on a ftand, and the bottom ftrewed with
fand, that they may duft themfelves to get rid of
troublefome infect.

When they are confined only to hemp-feed,
their plumage becomes almoft black.

At firft the young are fed with poppy-feed,
and bread moiftened, afterwards with fheep's
heart, or veal minced fmall, with egg boiled hard;
this may be mixed with oats, grits, flax-feed,
hemp-feed, and feveral other feeds.

In time they may be tamed fo as to ftand, and
eat on the table, or out of the hand; the form of the
hind claw prevents their perching on the finger.

THE

The TIT-LARK.

The two outside tail feathers are white on the outer webs, and there is a white line above each eye.

The tongue is divided at the tip into several threads.

It frequents the meadows in most parts of Europe, and feeds chiefly on worms, beetles, insects, and several kinds of grain; whilst it is eating it flirts its tail like a Water-wagtail.

This is one of the smallest of the Lark kind; the female builds her nest in low and marshy situations, lines it with horse-hair, and conceals it very artfully. She lays five or six eggs, and whilst the female sits, the cock perches himself upon a tree at a little distance; for Tit-larks can perch, the hind claw being bent, and therefore suited for that purpose; but he generally chuses a branch that is not very small, and depends most upon the claws of the toes before.

The cuckow frequently lays its egg in the Tit-lark's nest, the little cuckow has been found in it newly hatched, the old Tit-lark covering and feeding it, whilst the young Larks, though not quite feathered, were sitting on the outside of the nest. The Tit-lark is a timorous bird, and rises at the slightest noise; its song is plaintive, and

inter-

interesting, though perhaps not so varied as the Sky-lark's; it has something of the melancholy of the Nightingale's, but is not so continued, and the female sings in this, and perhaps in other species of the Lark, tho' there is reason to suppose that the song of the male bird is superior. Tit-larks sing as they sit on the ground, whilst perched upon trees, or sporting in the air, especially as they are descending.

THE WOOD-LARK.

Its head is surrounded with a pale band.

The first wing feather is shorter than the second.

The tail feathers dusky, about half of the two outer are marked obliquely with white.

There is a white wedge-shaped spot on the second, third, and fourth.

inhabits Europe, but not so far north as Sweden; and is found in Siberia, as far as Kamtschatka.

Wood-larks fly in flocks; they do not, like Sky-larks frequent open situations, but generally build their nests by the side of a wood, though they seldom take shelter there. The nest is made of dry grass, and lined with soft hair, the outside is moss; it is placed on the ground, the female lays five eggs; she builds very early, and breeds twice

in

in th year, the firſt brood are ſometimes ready to fly before the beginning of April.

The Wool-lark, like the Tit-lark, can perch upon trees, but for the ſame reaſon chuſes the larger branches, and like the Sky-lark it ſings, as it ſoars in the air.

It is a very tender bird, and preferved with great difficulty. Indeed it ſeldom live a year in a cage, or ſurvives moulting. Its ſong is very melodious, and plaintive, much like the Nightingale's, and like the Nightingale it ſometimes ſings in the night. Some have thought that its ſong has a reſemblance to the Blackbird's.

THE MEADOW-LARK.

The tail feathers duſky, the lower half white, except in the two in the middle; the throat and breaſt yellowiſh; over each eye a ſtripe of yellowiſh white.

It is not found farther north than Sweden.

Meadow-larks frequent brakes, and ſtubble lands, and after harveſt they are found in great numbers amongſt oat ſtubbles.

The male is rather larger than the female; in plumage they nearly reſemble each other, but the males are ſo quarrelſome that

they

they may eafily be known when they are brought
together, for they immediately begin to fight, with
every appearance of jealoufy and hatred.

In the fpring the male perches upon trees,
either to call or to difcover the female; fometimes too he rifes in the air and fings, but he
foon alights again on the ground and joins the
female.

The female, when any thing approaches her
neft, difcovers it by her fears, and her cries; in
this fhe differs from almoft every other fpecies of the Lark, for in general they fit ftill and
motionlefs.

The Meadow-larks are tamed with difficulty;
their wings muft be tied at firft, or a piece of filk
fhould be ftretched below the top of their cage,
to prevent their hurting themfelves againft the
wires. For fome time at firft they ought be to
kept in the dark, and the light fhould be admitted
by degrees.

THE GRASSHOPPER-LARK.

The tail feathers are dufky, the outer ones white half
way upwards, the next have a white wedge-fhaped fpot at
their points. There are two lines of white upon the
wings.

The

The Grafshopper-lark inhabits England, Germany, and Sweden: It is the smallest species, and in some parts of England is called the Pipit, from its cry. In the winter its note is thought to resemble the chirping of the Grafshopper, on which account it is likewise called the Grafshopper-lark. It makes this noise as it flies, or whilst it is perched upon a branch, for it can perch, and even upon small branches, its hind claw is better suited for this purpose than that of any other Lark, because it is not so long in proportion, and is less strait.

In the spring the male bird sings with a great deal of action, fluttering his wings, and raising his feathers; his song, though simple, is melodious, and the notes are distinctly pronounced. In the summer he chirps the whole night.

This species appears in England about the middle of September, many are taken about the neighbourhood of London: they flutter a little way rather than fly. They are very artful birds, skulking in the middle, and thickest part of hedges, and running along for a hundred yards together.

The female makes her nest in a very retired situation, but on the ground, so that the young

are

are often devoured by fnakes. She lays five eggs.

THE CRESTED-LARK.

The tail feathers are black, the two outer are white en the outer webs. The head is crefted.

This Lark is found in moft countries in Europe, but not fo frequently as the Sky-lark; it is moftly feen on the banks of rivers, by the fides of ditches, or running along the tops of furrows. In the winter it frequents the high roads to pick up any fcattered grain.

It perches fometimes on walls and roofs of houfes, the female builds her neft like the Sky-lark, but generally near the high road, and lays five eggs. She does not fit very conftantly, indeed the eggs do not feem to require fo great a degree of heat, or at leaft fo continued a warmth as thofe of fome other birds. The fong of the male bird is very melodious, and even loud, but at the fame time fo little difturbing, that a fick perfon might bear it without much inconvenience in his chamber. For the fake of their melody they are often confined in a cage. Though naturally they are not very wild, yet they cannot bear a long confinement; for this

reafon

reason it is usual to release them at the end of June, after which they seldom sing. It is easy to catch others the next spring, or indeed their song may be enjoyed during the winter, by placing a young Canary bird near them, who will easily learn it.

Their singing is attended with a tremulous motion in their wings. They are the first to announce the return of spring! If the weather be clear, and serene, they sing early in the morning as if to welcome the rising sun.

In gloomy weather they seem melancholy, and are silent.

The Crested Larks are often destroyed by birds of prey.

The CALANDRE-LARK.

The outer tail feathers are entirely white on the outer web, and about half way from the end of the inner; the second and third tail feathers are tipped with white.

On the breast is a black band almost in the form of a crescent.

They inhabit Italy, Sardinia, and the neighbourhood of Aleppo in Turkey in Asia; and the Deserts bordering on the Don, and the Wolga. They are likewise found in North Carolina.

The

The Calandre is one of the largeſt ſpecies of Larks, his beak is ſhort, and ſtrong, ſo that he can eaſily break many kinds of ſeeds. His ſong is louder than the Sky-lark's, but ſo melodious, that in Italy it is not unuſual to ſay of a good ſinger that he ſings like a Calandre. This bird too can eaſily acquire the notes of the Linnet, the Goldfinch, the Canary, and other birds, for it is an excellent imitator, and frequently mimicks the chirping of chickens, and the mewing of a cat. It is of ſo wild a diſpoſition when firſt confined, that it is neceſſary either to tie the wings, or to ſtrain a piece of ſilk acroſs the upper part of the cage; when it is a little accuſtomed to confinement it ſings almoſt continually.

THE SHORE-LARK.

The tail feathers are white on the inner half, the throat is yellow, and it has a black band paſſing under each eye, a little way down the neck; on the lower part of the neck is a broad black band.

It inhabits North America, Germany, Ruſſia, and Siberia. In winter Shore larks come in great flights to the ſouthern provinces of North America, frequenting ſand hills on the ſea ſhore. They feed on graſs ſeeds, and the buds of

the

RED LARK from PENSILVANIA

the birch tree, they run into fmall holes, and keep clofe to the ground.

They are efteemed delicate food, and frequently taken in fpringes made of horfe-hair, which are fet in fome place where the fnow is removed, and chaff is ftrewed about.

Some fay that it has little or no fong, and others fay that its fong is delightful.

GENUS 68.   STURNUS.

THE STARE.

The beak is flender, depreffed, a little blunt, the upper mandible very ftrait.

The noftrils guarded above by a prominent rim.

The tongue hard and cloven.

THE STARLING.

Its beak is yellowifh, the plumage black, gloffed with changeable blue, purple, and copper colour; with white or pale yellow fpots.

It inhabits both Furope, and Africa; for it is a bird of paffage, migrating from one country to another, and is found from Sweden, to the Cape of Good Hope.

The Starling breeds in hollow trees, fometimes
feizing

feizing the neft of the Wood Peeker, who in its turn alfo feizes upon the Starling's neft. Sometimes too it will ufe the old neft of a Thrufh, for it is not nice in thefe refpects, collecting only a few leaves, or other dry materials to lay its eggs upon; occafionally they breed under the eaves of houfes, in towers, ruins, in cliffs or high rocks over the fea; the neft is frequently made of ftraw and fmall fibres of roots, and lined with leaves.

Starlings are gregarious birds.—In winter they affemble in flocks. They collect in immenfe numbers in the fens of Lincolnfhire, and do great damage among the reeds, by roofting upon them in fuch multitudes, as to break them down by their weight; for reeds are ufed there for thatch, and are the harveft of that country; fometimes they are feen in company with Redwings, and Fieldfares; they are eafily diftinguifhed from other birds by their manner of flying, which is very irregular, and diforderly.

They feed principally on worms, and infects; they will fometimes eat grain, and fome kinds of fruits, and they have been known to get into Pigeon-houfes, to fuck the Pigeon's eggs.

They are very docile, and difpofed to imitate;

for

for this reason they are often confined in cages, and taught to speak.

Their flesh is very bitter.

Of this species there are several varieties, one that is wholly white, the bill reddish, and the legs flesh coloured.

Another variety that is white, spotted with blue.

A third is cream-coloured, with white spots.

There are several other species of the Stare, but nothing remarkable is known of their manners, except the

WATER OUZEL,

Whose plumage is black with a white breast.

It inhabits Europe, frequenting watery places; in the winter it is generally seen near cataracts, or falls of water, and springs that do not freeze; though it is not web-footed, it descends in a wonderful manner through rapid whirlpools, to feed upon the onisci, and other insects, that are at the bottom; and walks about upon the ground under the water, in search of this kind of food; it devours insects, and small fish only, and does not feed on grain; it frequently wags its tail, and its nostrils are almost covered with a little rim.

It

It breeds in holes in the banks, and lays five white eggs tinged with red; the nest is curiously made of hay, and fibres of roots, and lined with dead oak leaves, having an entrance of green moss.

GENUS 69.   TURDUS.

THE THRUSH.

The beak in this Genus is straitish, the upper mandible a little bending towards the point, and slightly notched.

The nostrils are naked, in some species they are half covered above with a little membrane or skin.

The corners of the mouth are furnished with a few slender hairs.

The tongue is a little jagged at the end.

The outmost toe adheres as far as the first joint to the middle toe.

The first species is the MISSEL-THRUSH.

Its back is dusky, its neck and cheeks are white, or yellowish white, spotted with brown; the base of the lower mandible, and the gape, yellow, the other parts of the mandible dusky.

It inhabits Europe, in England it stays all the year, but in some parts it is a bird of passage.

The Missel birds feed on berries, of various kinds; the holly, ivy, hawthorn, and particularly

the

the berries of the missletoe; indeed it is a means of propagating that plant, for the seeds passing through the body, and being let fall on trees, grow there.

They build their nests in bushes, or low trees, or by the side of a tree, frequently the ash tree; it is made on the outside with twigs, then with moss, and leaves, and lined with fine withered grass. The hen lays four or five eggs, of a dingy flesh colour, marked with reddish spots.

The Missel bird's song is very agreeable, though not so melodious as that of the common Thrush; they sing early in the spring, from the summit, or top of a very high tree, from which they may be heard to a great distance, varying their note continually; sometimes too they sing early in the year, when the weather is blowing, and stormy; on which account they are called storm cocks.

They are good eating. Some varieties of this bird have been found, one of a reddish cream colour, the stomach white with cream colour spots; another variety white, spotted with brown beneath.

The FIELDFARE is another Species.

The tail feathers are black, except the outer ones, which are whitish at their points on their inner margin. The head and extreme part of the body grey, or a dusky white.

Fieldfares are found all the year in Poland and in Sweden, in prodigious flocks; in England and France they are only birds of passage, making their appearance in winter, or the latter end of autumn, and returning in the spring. They appear constantly in the Orkney Islands, at the approach of winter, in their passage. Their food is berries; in Sweden they frequent those parts where the juniper grows, and there they build in high trees.

Their flesh is esteemed. The Romans kept these birds, and the Redwing, by thousands in their aviaries to fatten, esteeming them a delicacy.

The REDWING, the 3d Species of this Genus.

The wings are rust coloured beneath, its eyebrows are whitish.

It visits England with the Fieldfare, but with us has only a piping note. In Sweden it warbles melodiously in the spring, perched on the top of a high tree, in Maple forests. It builds in some

4   low

'ow fhrub or hedge, and the female lays fix blueifh green eггs, fpotted with black.

Redwings, common Thrufhes, and Fieldfares, abound fo much, and are killed in fuch numbers for the markets in Polifh Pruffia, that a tаx has been paid for 30,000 at Dantzic only, in one feafon.

The THROSTLE.

The wing feathers are ruft-coloured at the bafe, or the infide; the under part of its body is fpotted, the fpots are fhaped like the heads of arrows, with their points upward.

This is one of our fineft fong birds, and fings the longeft. It begins its melodious warbling early in the fpring, and continues until the end of autumn. Perched upon fome high tree, it amufes us in our walks with its delightful fong.

With us the Thrufh is not gregarious, and though it remain in England all the year, poffibly it may go from one part to another, at the approach of Winter, or retire into the woods. In France Thrufhes are migratory, appearing in Burgundy twice a year: firft in April, and leaving it in May; and again at the vintage, when

when they do a great deal of damage in the vineyards, by eating the ripe grapes; they difappear
again at the approach of winter.

The REDBREASTED THRUSH.

The upper parts of the body olive brown, the
is reddifh, the eyelids are white, and it has a white fpot
each fide between the beak and the eyes: the tail is black,
edged with olive brown.

This is a North American bird, and retires from
the warmer into the more northern parts to breed.
In fome climates Red-breafted Thrufhes build their
nefts, lay, and hatch their eggs, in fourteen days:
in colder fituations they require twenty-fix days.
The neft is compofed of roots and mofs, and the
hen lays five eggs: both male and female fhare
the care of making the neft, and feeding the
young; whilft the female is fitting, the male frequently cheers her with his mufical voice.

Though they migrate to colder climates in the
fpring, this does not appear to arife from their being
unable to fupport the heat; becaufe there has
been an inftance of a fingle bird of this fpecies remaining all the fummer in Virginia, feeding on
the berries of a tree, which had been lately introduced into the country, and of which it feemed
very fond.

z   Their

is chiefly worms and insects; they
the feed of fall fruits, or turnip and
and of these ... they are so fond,
... flesh, at the end of the year,
... tinged of a purplish colour.

... these birds is very ... ... in their
... they will not bear confinement in
... they are not shy or distrustful, for
quently seen hopping on the ground
...

... MIMIC THRUSH,

... ash-colour above, beneath of a ... ...
... feathers black, on the ... ...

... Virginia, and Carolina, and indeed
of South America: it sings most me-
... ... from shrub to shrub, and imi-
... of almost every animal it hears;
... Nightingale does not exceed its
... that it sings the greatest part
... It frequents moist woods, and
... both ... and trees, but it ... so far,
... ... with ... it will ... ... its
... ... or if ...
when taken, are reared with great

C

The MOCKING THRUSH.

Its back is dusky, its breast and two outmost tail feathers are white, and its eyebrows are white.

It is an inhabitant of America.

Suspended in the air, it enraptures those who hear it with its song; which no melody can rival.

Though its natural notes be inexpressibly sweet, its powers of imitation are such, that it can mimic the song of every bird it hears, and any other noise however discordant; the mewing of a cat, the chattering of a magpye, or the croaking of a frog; from the variety of its note, it has a name among the Mexicans which signifies 400 tongues. It begins generally with its own natural song, and introduces into it a variety of melody, which it has adopted from other birds.

In the warmer parts of America it is singing almost incessantly, day and night, from March, to August; and as though animated by its own inimitable song, it gradually raises itself on the place it is perched on, extending its neck, drawing it back, stooping, rising, fluttering its wings, and turning itself round with a variety of action.

The

The Mocking Thrush has great courage, and is not afraid of attacking birds much superior in size.

It is however very tender, and found in those parts of America which are to the north of Carolina, only in the summer season; in the more southern climates it continues the whole year, building its nest in June; the hen lays six eggs.

It often builds in fruit trees, and frequents the neighbourhood of houses; it is a familiar bird, often finging melodiously, perched on a chimney top; yet it will forsake its eggs on perceiving that its nest is discovered.

The STONE THRUSH.

The head is blue, the tail ruft colour, the body reddish, with small dusky spots; the two middle tail feathers are dusky, the rest are reddish, except at the very tips.

It is found on the mountains of Switzerland, of Austria, of Tyrol, and the Alps.

This bird receives its name from the places which it frequents, for it generally stands upon a stone; it appears to be very diftruftful and suspicious, and seems careful to keep out of the reach of a gun, for upon any perfon's approaching within gun fhot, it immediately flies away, alighting again upon a ftone, fo that it can fee all around.

The female builds her neft in the top of a cavern, or the hole of a rock, difficult of accefs, and lays three or four eggs.

The neft is artfully concealed, and when after great toil, and fome danger, you have found it, the old birds defend it with great refolution, ftriking at the eyes of thofe who attempt to invade it.

The young ones muft be taken before they can fly, for they are fo artful, and fo cautious, that they cannot be enfnared in any traps.

Though their natural food be worms, and infects, when brought up in a cage they will live upon the fame food that is given to young Nightingales.

Their fong is very melodious, and they eafily learn the notes of other birds, and even tunes. They generally fing at day break, and at fun-fet; and if a candle be brought to their cage in the night, they will begin their fong. In the day time they often feem to be practifing their notes in a lower key.

THE ROSE-COLOURED OUZEL.

The plumage is of a faint rose colour, the head, wings, nd tail, black; and the back part of the head crested.

This is a beautiful, and indeed rather a scarce bird in Europe; two or three have been found in England, and a few near Bolga in France, and feveral in Burgundy: the back feathers on the head, wings, and tail, are gloffed with a beautiful caft of green, and of purple; the breaft, the ftomach, the back, and the fmaller coverts, are tinged with a delicate rofe colour, of two different fhades.

Little is known of the habits of this very elegant bird; great flocks appear every year near the river Don, and the Irtis, and on the borders of the Cafpian Sea, they devour great numbers of locufts, and breed amongft the rocks. In the fummer they are found near Aleppo, where they are called the Locuft bird.

THE BLACKBIRD.

Its body is black, the cak and eye brows yellow.

It inhabits Europe, frequents woods, and difeminates juniper.

The

The beak is not emarginated or notched, the female dusky rather than black.

The Blackbird does not, in many inftances at leaft, migrate from one country to another, but in winter conceals itfelf in hedges, and thickets, in very thick woods near fprings, that do not freeze; efpecially where there are evergreens, that it may be the lefs expofed, and where perhaps it can beft fupport itfelf with food during that rigorous feafon.

Blackbirds fometimes vifit our gardens in fummer, and are great devourers of fruit. In a wild ftate they live on almoft all kinds of berries, fruit, and infects: when confined they are fed in part with meat, boiled, and minced; they will eat bread and butter, and are oftentimes kept on a kind of pafte, which may be preferved for a long time without fpoiling.

The kernels of the pomegranate are poifonous both to them, and to Thrufhes.

They are very delicate food, and in thofe countries where they feed on olives, and myrtle berries, their flefh acquires a juicinefs, and a perfume, which is very agreeable.

They are perfecuted both by men, and birds of prey, which prevents their increafing very faft;
this

BLACK BIRD.

his otherwife woul1 probably be the cafe, for they begin to lay very early in the fpring, and lay more than once. Their neft is a little like the Thrufh's, but lined within; the outfide is compofed of mofs, and mud, and lined with bents of grafs, and fibres of roots; and they are fo induf rious, that it is finifhed in eight days. They generally build low, in bufhes, or in trees that are not very high, and there have been inftances of their making their nefts in the hollow of a tree. Mr. De Salerne fays, that a Blackbird having built its neft in a hedge, and very near the bottom, finding that two hatches of young had been deftroyed by cats, the third time placed it in an apple tree, eight feet above the ground. The female lays five or fix bluifh eggs; the male does not affift her in fitting, but provides her with food.

A gentleman fays, that he had feen a blackbird that was hatched early in the year, undertake the care of fome little ones that were lately hatched.

Although Blackbirds lay more than once in a feafon; their firft laying frequently mifcarries, from the inclemency of the weather, in the early part of the year.

The young birds are fuppofed to moult more than once the firft year, each time the feathers of the cock bird which at firft are of a dufky reddifh brown, like those of the hen, become more black, and the sides of their bills more yellow.

The old birds moult at the end of the summer; they begin to sing early in the spring, and sometimes again after moulting.

They do not fly in flocks, but having a very quick sight, and being naturally solitary and timid, they do not suffer themselves to be approached: Yet they are not very sullen when taken, and are easily tamed: In confinement they may be taught other notes than their own; to whistle, and even to speak.

Their note is very fine, but too loud to please, except in the open air.

They are taken easily in traps. Being naturally restless, when confined, they are teazing to other birds that are shut up with them: therefore they are improper birds to be kept with others in an aviary.

A blackbird is supposed to live about seven years. There have been instances of an accidental variety of this species, whose feathers were mostly white.

THE RING OUZEL.

Its plumage at a distance seems blackish, but upon nearer view each feather is margined with grey, or ash colour; on the breast is a large patch, or crescent of white, and the inside of the beak is yellow.

The Ring Ouzel seems to be a bird of passage, though his migrations are not very distinctly known. In October they are seen in some parts of France, they come there in small flocks, from twelve to fifteen, and stay but a few weeks : They disappear when the frosts begin, and return again in April or May. At certain seasons they are seen in greater numbers in Sweden, Scotland, Savoy, Switzerland, and Greece : it is probable too that they are dispersed in some parts of Asia, Africa, and in some of the islands near the African coast.

They have been observed in Hampshire twice a year, in April, and September, in flocks of twenty or thirty; those that breed in Scotland, and Wales, never leave their native country: in Scotland they breed amongst the hills, but come down to feed on the berries of the mountain ash.

On Dartmoor in Devonshire, they settle in

the

the banks, by the fide of ftreams, and are very noify when difturbed; their nefts are placed at the bottom of bufhes clofe to the ground, conftructed very much like the Blackbird's, and they lay five eggs.

They feed on berries, and infects, they are fond of ivy berries, and of grapes; in France, during the vintage, they get very fat, and their flefh is then rich, and juicy. They feed too upon fnails, and are very dextrous in breaking their fhells on a ftone, to enable them to get at their contents.

The REED THRUSH.

The plumage on the upper part of the body is of a reddifh brown, beneath of a dingy white; the wing feathers edged with a reddifh brown.

It inhabits Europe, particularly the marfhy parts, where reeds grow, and climbs up them in the fame manner as Woodpeckers climb trees; it builds its neft fufpended from three reeds, faftened together, or on mofly hillocks amongft the reeds and rufhes. The male fings continually, whilft the female fits; from which it has been called by fome the Water Nightingale.

The CHANTING THRUSH is found in the fouthern provinces of China; it is there called the Nightingale.

BLUE THRUSH.

Nightingale, and is said to be alm st the only bird in that extensive empire that has a song.

The PIGEON THRUSH is found in the Philippine islands; like the Starlings in Europe, it often builds in pigeon-houses, from which it has its name.

The GREEN THRUSH comes from China, it sings well, and like the Starling is fond of wetting itself with water.

The GREY THRUSH inhabits the coast of Coromandel; it is mostly hopping on the ground after insects, which it finds there.

The YELLOW CROWN'D THRUSH inhabits Ceylon, and Java; it is frequently tamed, and kept in cages, and is very imitative.

The NUN THRUSH inhabits the woods of Abyssinia, feeding on berries and fruits, frequenting such trees as are upon the brinks of precipices, which makes it difficult to be shot, or at least to be procured when killed.

The

The BLACK-BREASTED THRUSH inhabits Cayenne; the throat and breast are black, bordered all round with white like a cravat.

The MUSICIAN THRUSH, is black, mixed with brown; the chin, under the eyes and throat, are reddish orange; and on each side the neck, is a broad patch of black, spotted with white.

It inhabits Cayenne, and seems of a solitary and timid character; seldom quitting the trees where it perches, but for food. Its song is very melodious, and thought to resemble a flute; by many it is esteemed superior to that of the Nightingale, from the natives this has procured the bird its name of the Musician. It feeds on ants and other insects.

In Cayenne is found the CHIMING THRUSH; this species generally flies in small flocks, of about half a dozen; they have a singular cry, resembling a chime of three bells, of different notes, from which they take their name.

Another kind of Thrush found in Cayenne has a remarkable cry, very loud, and like the alarum of a clock, which it repeats morning and evening;

evening; on this account, it is call'd the ALARUI THRUSH.

There is a species of Thrush call'd the Ant-eater, which likewise inhabits Cayenne; it fee's on ants, and runs up trees like the Woodpecker, supporting itself in the same manner by its tail.

GENUS 70.   AMPELIS.

The beak is strait, and convex, the upper man ble the longer, and slightly bowed, both mandibles are toothed, or notched near the end, the nostrils are hid in bristles.

The tongue is sharp, cartilaginous, and divided.

The middle toe connected at its origin to the outmost toe.

There are no birds more distinguishable than those of this genus, for the delicacy and variety of their colours, or the beauty and gloßiness of their plumage.

All the richest colours of nature, the various shades of violet, purple, blue, orange, red, pure white, and velvet black, are lavished here with

5   gayest

gayeft profufion; fometimes moft delicately foftened into each other, and fom times moft ftrikingly
contrafted, and the changeable fhades which the
fame feathers exhibit in different points of view,
produce a fplendour of effect, which words cannot exprefs, the imagination conceive, or the pen-
cil imitate.

All the fpecies of this beautiful genus of birds
inhabit only the warmer climates of America,
feldom more to the fouth than Brazil, or more
northward than Mexico.

In this range of country they feem migratory;
though they appear nearly at the fame time, they
do not affociate together; but frequent marfhy
fituations, finding upon the vegetables which
grow there, the infects on which they feed.
They are of all fizes, from that of a Pigeon to
a Thrufh. Many are deftroyed both for the
fake of their feathers, which are beautiful, as well
as for their flefh which is delicate food. They are
fuppofed to be very deftructive to the rice plantations; if that be cafe, it is no wonder that the
natives fhould endeavour to prevent their increafing too faft.

The firſt ſpecies is the WAXEN CHATTERER.

The back part of the head is ſlightly creſted; the ſecondary feathers in the wing are covered at the tips, with a ſealing-wax-like ſubſtance.

The birds of this genus inhabit Europe, Aſia, and North America. It is difficult to ſay what parts of Europe the Chatterers principally frequent, becauſe they are occaſionally found in moſt; a very few in England. About Edinburgh they appear annually in the month of February; they are numerous in Germany, and perhaps in greater numbers ſtill in the northern parts of Europe, for they ſeem rather to paſs through, than to migrate into the ſouthern parts: they have been ſeen to arrive in Italy in flocks of more than one hundred, and they have paſſed through Sweden in ſuch numbers as almoſt to intercept the light. It is not eaſy to determine what occaſions their leaving their native country, which perhaps may be Tartary.

Their migrations take place ſometimes once in three, and ſometimes once in ſeven years, and may ariſe probably from an occaſional ſcarcity of food.

They

They are particularly fond of grapes, they
on the berries of the mount. in ash, the white
thorn, the eglantine, the laurel, almonds, appl
fervices, wild goofeberries, figs, and almost a
juicy fruits.

They are fuppofed to breed in Tartary, and to
build their nefts in the holes of rocks.

They are of a gentle difpofition, but can only
bear confinement for a fhort time. One that
was kept in a cage was filent all the while; it
would never eat any grain, but frequently drank,
dipping its beak in the water eight or ten times
fucceffively.

They are very neat, always removing every
thing that is difagreeable to one corner of the
cage.

The birds of this fpecies are of a focial difpofition; they fly in flocks and feem much attached. The male and female carefs, and
feed one another, and the fame affectionate attentions feem to fubfift amongft the males towards
the males, and likewife amongft the females.

The Chatterer is very delicate food, its fong is
faid to be pleafing, though it feems very little
known.

Another

Another Chatterer, a variety perhaps of this species, is found in America. It has been observed that the breastbone of this bird is advantageously formed for cutting the air, from which it seems intended for distant migrations. It's found from Canada to South Carolina. The female has not the scaling-wax-like substance on her wing.

The POMPADOUR CHATTERER.

Its plumage is of a beautiful purple, the greater coverts of the wing are long, narrow and hollowed beneath, the tips without webs, hanging elegantly over the quills.

The birds of this species inhabit Cayenne in North America; are migratory, appearing in Guiana in March and September, the seasons when those fruits are ripe on which they feed; they frequent high trees by the water-side, and build their nests on the highest branches. They never retire into the interior parts of the woods.

There are several other species of this beautiful genus inhabiting nearly the same countries, and probably resembling those which have been described in their manners; we will however mention one other species called the

C A₄

CARUNCULATED CHATTERER,

It is about twelve inches in length, the plumage a pure white, the bill black, an inch and half in length; at the bafe is a flefhy fubftance, like that of the turkey, except that it is flaccid, and pendulous, when the bird is in a flate of repofe, but when it is agitated, this flefhy fubftance fwells, lengthens to the extent of two inches, and becomes erect in confequence of the air which is forced into it through the opening of the palate. The cry of the Carunculated Chatterer is very loud, and feems compofed merely of two fyllables, *ang*, *on*, pronounced in a drawling manner. It inhabits Brazil and Cayenne.

GENUS 71.   LOXIA.

The beak is convex above and below, ftrong and very thick at the bafe.

The lower mandible inflected, with a lateral margin.

The tongue, as if cut off at the end.

THE COMMON CROSS BILL.

The two mandibles crofs each other different ways, at the point; but it is very fingular, that in fome the lower mandible croffes to the left, and

in

... (....rs to the right; however awkward this conformation may appear at the first view, we shall upon reflection find, that it is judiciously adapted to their way of life, and very well contrived to separate the scales of the fir cones, and to take out the seeds, upon which they principally feed. They place the lower mandible under the scale, and separate it with the upper; it is remarked that they seldom break the cones in the trees on which they grow; by this means they disseminate or plant firs.

The beak is also very useful in assisting them to climb; they use it for this purpose, almost in the same manner as Parrots. With one stroke of the bill they can split an apple; this they frequently do in order to get at the pippins, the only part of the apple on which they feed; in confinement they will feed very well on hemp-seed.

This bird inhabits those countries only where fir trees grow, and principally cold climates, or the mountainous parts of such as are temperate.

It is found in Sweden, Poland, Germany, Switzerland, in the Alps, and the Pyrenean mountains; in these countries it resides throughout the year. Sometimes the Crosbills migrate in great flocks, and are found in England, and other countries;

countries; but these migrations are suppofed not to be conftant and regular, and may perhaps proceed from an accidental fcarcity of that food of which they are fond, in thefe countries which they generally inhabit; or perhaps they may have been carried away by fome ftorm of wind, for they have been obferved to arrive in great numbers, and fometimes fo much fatigued, as to be eafily caught by the hand: a Crofsbill has been found in Greenland, but it was probably accidentally driven there, as it could not in that dreary region have found its proper food.

Crofsbills in this refpect are faid to differ from other birds, that they make their nefts in the depth of winter, in the month of January, on the higher branches of fir trees; faftening them with the refin of the fir, and plaftering them in fuch a manner with that fubftance, that the wet from the fnow, or rain, can never penetrate. The corners of the mouth of their young, as in moft other young birds, are yellow, and it is probable that the hen lays four or five eggs.

The colours of the plumage of thefe birds differ, in different individuals, fcarcely any two being exactly alike; and the colours in the fame bird frequently vary, from a dark blackifh fhade, to fhades inclining to red, or green.

The

The plumage of the female is less bright than that of the male. It seems to be a very stupid bird, suffering itself to be approached, sometimes so as to be killed with a stick. It scarcely flies away when it is shot at, and being as inactive, as it is unsuspicious, it is often destroyed by birds of prey. In confinement it is silent, t... in a natural state its notes are very fine; it is easily tamed, and is good food.

This bird is very frequent in the fir woods near Bath, and at Barr in Staffordshire at certain seasons of the year.

The GROSBEAK or HAWFINCH.

Its first quill feathers have each a spot of white about the middle of the inner web; the four outer quill feathers, seem to be cut off at the tips in the form of a bittle a ..., and are bent at the end.

The tail is black, the two middle feathers incline, to ash colour, and all the other feathers have the one half white on the inner webs, and tips.

It inhabits the more southern parts of Europe, and feeds upon the kernels of cherries, which it discriminates, it is injurious to gardens.

The Hawfinch is found in all the temperate parts of Europe, from Spain, and Italy, to Sweden.

In France, when the winters are very severe,
these

thefe birds difappear for a fhort time. In the fum
mer they frequent woods; and orchards in the winter. In Burgundy there are fewer in winter than in
fummer; they feem to arrive in numbers in April,
they fly in fmall flocks, and perch in copfes. They
build in trees about ten or twelve feet from the
ground; like the neft of turtles, theirs are compofed of twigs of dry wood, faftened with fmall
roots; they lay five eggs, blue, fpotted with
brown: as their numbers are not confiderable, it
is probable that they breed but once a year. They
feed their young on infects, and chryfalis's; and
if any attempt be made to rob their nefts, they
defend them with great refolution with their
beaks, which being ftrong enough to break cherry-
ftones; enable them to bite very hard.

Befides fruit they eat many infects, but when
confined they conftantly refufe flefh. In their
natural ftate they feed on almoft all kind of grain,
and on the kernels of fruit; on the feeds of
firs, the maft of beech, on walnuts, and almonds.

Their hearing feems very imperfect, and they
have very little note. They are not taken with
a bird call.

When they are confined, it is proper to put
them in cages, feparate from other birds, for
without appearing at all irritated, they are very
apt

apt to deſtroy birds ſmaller than themſelves, by pinching out pieces of their fleſh with their bills.

They are migratory, and ſeen in England only occaſionally.

The PINE GROSBEAK

Has two bars of white on the wings. On the head, neck, breaſt, and near the tail, the plumage is of a roſe coloured crimſon.

It inhabits foreſts of fir trees in Sweden, and Canada, and the pine foreſts of Invercauld in Aberdeenſhire, feeding upon the ſeeds in the cones, and diſſeminating the firs.

In the night it ſings very melodiouſly, and almoſt without ceaſing.

In the winter it migrates into the more ſouthern provinces of Sweden. The plumage of the younger inclines to a reddiſh, in the older birds to a yellowiſh caſt.

It makes its neſt on trees, at a ſmall height from the ground, with ſmall twigs and fibres, lined with feathers. The female lays four white eggs, which are hatched in June. The Pine Groſbeaks are taken about Peterſburgh in autumn, in great numbers, and brought to market; in ſpring they retire to Lapland.

THE

THE BULLFINCH.

Its legs, wings, and tail are black; the lowest feathers of the tail, and the hinder wing feathers, white.

It inhabits the woods in Europe, and disseminates the service trees. The male is red underneath, the female dusky ash colour. They are frequent in Russia and Siberia, and caught for the table.

It may be taught to whistle any tune, and seldom forgets what it has learnt; it will become so tame as to come when called, perch on its master's shoulder, and whistle a tune when commanded.

There have been several instances of their becoming quite black.

The female Bullfinch may be taught to pipe as well as the male. It has been said, that a gentleman who had whistled some tunes to a Bullfinch, was very agreably surprised to hear the Bullfinch imitate the song, but introducing at the same time some elegant variations, which were so well executed, that the teacher acknowledged himself to be surpassed: Though on the other hand it has been observed, that a Bullfinch who had only heard some ploughmen whistle, learnt

to

whistle in the same loud, and uncouth manner.

This species seems to be of an affectionate character, and capable of attachment: one which had been tamed, escaped from its cage, and lived in a wild state for more than a year; upon hearing the voice of the girl who had taken care of it, it returned to her again. It is related that when they have been taken from those who have been accustomed to feed them, that several have been known to pine away, and die of grief. There is a story told of one that is hardly to be believed; his cage is said to have been thrown down by some dirty and ragged people; the bird did not seem very much hurt at the time, but it never saw any mean, shabby person afterwards without being seized with convulsions, and at length died in one of these fits, about eight months after the accident.

The following instance is related of their affection to one another: four young Bulfinch's of the same nest were reared together, the three strongest, as soon as they could feed themselves, assisted in feeding the smallest, who was not yet capable of providing for itself.

The male and female, after the breeding season is over, continue to fly together.

PART VI.   D   The

The Bullfinch is not uncommon in England, and makes its nest in bushes; it is composed chiefly of moss; the eggs are of a blueish white, marked at the larger end with dark spots. They breed the latter end of May, and beginning of June.

In summer the bull-finch mostly frequents woods, and retired situations; in winter, it haunts gardens and orchards, and is very destructive to the young buds. In Germany great pains are taken to teach these birds to pipe, and even to speak, and many are imported therefrom.

THE CARDINAL GROSBEAK.

It has a beautiful red crest, the beak is of a pale red colour; round the beak and on the throat it is black; the rest of the plumage is a rich scarlet.

It inhabits North America, from Newfoundland to Louisiana, and feeds upon maize, buckwheat, &c. of these it frequently collects a hoard, perhaps equal in quantity to a bushel, and artfully conceals its magazine, with leaves, and sticks, leaving only a small entrance. The colours of the male are more beautiful, and vivid, than those of the female. It is a hardy and vigo-
rous

rous bird, eafily tamed, and its natural notes are very pleafing.

Indeed it is often called the Virginian Nightingale. They warble their melodious fongs in the mornings of the fpring, perched on the higheft trees. They will fing in a cage, and are frequently brought to England; in a ftate of confinement, the male and female feem to harbour fuch a refentment, that they will often kill one another.

They feldom are feen in greater numbers than three or four. When tame, they will learn to whiftle; but their natural note is faid not to be unlike that of the Englifh Thrufh. In fummer they haunt the fwamps, in autumn retire to the fouth.

The JAVA GROSBEAK or JAVA SPARROW.

The plumage is dufky, the temples white, the beak red.

It inhabits Afia, and Ethiopia, frequenting plantations of rice, to which it is very deftructive.

The plumage of this bird is exquifitely delicate, both from the harmony of the colours, and becaufe it is fo fmooth, that it appears like velvet, or rather like that fine powdery bloom which is found on fome plumbs.

D 2   On

This bird is often defcribed on Chinefe paper.

The GREEN GROSBEAK, or GREENFINCH.

The plumage is of a yellowifh green, the quill feathers are edged with yellow, the four outer tail feathers are yellow from the middle to the bafe.

It inhabits Europe, is not frequent in Ruffia, and fcarcely found in Siberia. In the winter it frequents woods, fheltering itfelf from the intemperance of the weather, upon ever-greens, or amongft the fading foliage of the oaks.

It makes its neft in thofe trees, and fometimes in bufhes; the outfide is compofed of hay, and ftubble, the middle of mofs, the infide of feathers, wool and hair.

The female lays five or fix eggs, and fits fo clofely, that fometimes fhe may be taken on the neft with her young. The male occafionally relieves his partner, fitting upon the eggs in his turn; and amufes himfelf at other times, by flying in little circles round the neft.

Greenfinches are eafily tamed, fometimes they will eat out of the hand in five minutes after they are taken. To effect this, let it be carried into a dark place, and perch it upon your finger, it

will

will not attempt to move, not knowing the
count of the darkness wh...   to fly; bri...t...
the finger of the other hand under its br...,
raising the bird a little; from the uneasiness of
its situation on the finger, on which it was first
placed, it will climb on the other; if this be often
repeated, and the bird be stroked and caressed,
it begins to lose its fears; and if the light let
in by degrees, it will sometimes immediately eat
bruised seed out of the hand, and from that time
become tame.

There are many other species of Grosbeaks.

The Grenadier Grosbeak inhabits St. Helena,
and the Cape of Good Hope; frequenting watry places, and making a curious nest among the
reeds; it is constructed with small twigs, curiously interwoven with cotton, and divided into
two apartments, with one entrance; the upper
apartment for the male, and the lower for the female. The nest is of so close a contexture, as to
be impenetrable to the weather. The colour
of this bird is a brilliant red; amongst the green
reeds they appear like so many scarlet lillies, which
produces a very striking effect.

D 3   The

The Fantail Grosbeak inhabits Virginia, and carries its tail spread in an horizontal direction.

The Philippine Grosbeak makes a most curious nest, in the form of a cylinder, only swelling out in the middle. It is composed of the fine fibres of leaves, and suspended from the extreme branch of a tree. The entrance is at the bottom, and after ascending to the widest part, the proper nest is made there, towards one side; here the ingenious little bird hatches her young in security.

Another Grosbeak very nearly resembling the Philippine Grosbeak, makes a curious nest in a spiral shape, like the shell of the Nautilus; and suspends it from a slender twig, over the water, always turning the opening towards that quarter, where the least rain may be expected.

The Abyssinian Grosbeak's nest is not less ingeniously contrived; it is suspended like the others; its form is that of a pyramid, with an opening towards the east; the cavity is divided by a partition; after ascending a little way, the bird descends again into its nest, on the other side of the partition, which by this means is secured from monkies, squirrels, and snakes, as well as from

the

i' rd), which in the country contains r
months together.

The Penfile Grofbeak, which inhabits Madagascar, compofes its neft of ftraw and reeds, interwoven in the fhape of a bag, faftened above to the twig of a tree, which overhangs the water. The entrance is from below, and the real neft is placed on one fide, within this curious contrivance. Every year the bird faftens a new neft to the end of the laft, and five are often found hanging one from another. Like Rooks they build in fociety, and five or fix hundred nefts have been feen on one tree.

The Minute Grofbeak is only the fize of a Wren, it inhabits Surinam, and Cayenne. They live in pairs, the male never forfaking the female. It is a fprightly bird, frequenting lands that have been for fome time uncultivated, and even approaching habitations. The neft is about two inches acrofs.

Another fpecies called the Coly, is placed by Linnæus among the Grofbeaks, its tail is wedge-form, like the tail of a Pheafant.

D 4   G E-

Genus 72.   EMBERIZA.

The bill is ftrong, and conic, the fides of each mandible bending inwards; in the roof of the upper mandible is
a hard knob, well formed for the purpofes of breaking, and
bruifing hard feeds.

THE SNOW-BUNTING,

Like fome other birds that inhabit cold climates, has a fummer and a winter drefs; in the
fummer the plumage is tawney, but at the approach
of winter, the head, the neck, the ftomach,
and great part of the wings, become white.
Linnæus fays, they vary according to age and
feafon. They inhabit Greenland, Hudfon's Bay,
Spitzbergen, and the mountains in Lapland.
They feem to make the colder regions their fummer refidence, and it is wonderful when we confider that they are graminivorous, at leaft when
they are with us, how they can fubfift in Spitfbergen, where there is fcarcely any vegetation,
except of moffes, and plants of that clafs.

In the winter they leave thefe dreary regions,
and come to warmer climates, in amazing numbers; they fill the roads and fields in Sweden.

In

In the winter 1778, they came in such multitudes to one of the Orkney islands, as almost to cover it, yet scarcely any two agreed exactly in colour: they make the Feroe islands, Shetland, and the Orknies their resting places; they abound in the highlands in Scotland; their flocks are so great, and they fly so close together, that the fowlers destroy great numbers. When they first arrive, they are very lean, but soon become fat, and are a delicious food. In Scotland they are called Snow Flakes, and a few breed like the Ptarmigans on the highest hills. In their summer feathers, they are often seen in the south of England; now and then a milk white one is found. In Russia and Sweden, the flocks are immense; they frequent villages, some are white, some speckled, some brown, and even their winter dress seems as various, as their summer plumage; they are found also in Germany, where they are caught, and fed with millet, and are as delicious food as the Ortolan.

In the spring they are seen in vast flocks in Norway, and stay about three weeks: they then return to the dreary northern regions, in order to breed in security. They arrive in Greenland about April, build their nests in the clefts of rocks, in the mountains, in May; the outside is grass, it is lined with feathers, and the fur of the

arctic

arctic fox; the female lays five eggs, the male
fings melodioufly near the neft.

When they collect together in autumn, on the
fhores, in order to migrate, they are caught in
numbers and dried, by the Laplanders, who take
them in hair fpringes.

The ORTOLAN BUNTING.

The quill feathers are black; the three firſt feathers in
the tail are whitiſh on their margin, the reſt black, the inner
ends of the outmoſt feathers are marked with a great ſpot
of white.

This fpecies is migratory, and not found very
far to the north; their winter refidence is not
very well known; they appear with the Swallows in fpring, and fing very pleafingly in the
night.

Their neft is conftructed without much art,
they lay four or five eggs, and breed twice in the
fummer. When they firft arrive, they are very
lean. By epicures they are efteemed a moft delicious food. In order to be fattened they are
confined in a room which is lighted by lanthorns;
it is fuppofed that this uniform light, keeps them
quiet, and that they fatten the better; they are
fed with oats, and millet, until they become one

lump

lump of fat, and if they were not killed juſt at the proper time, they would die.

The Romans had aviaries built with great magnificence, and furniſhed with every convenience for the purpoſe of feeding Ortolans, and other birds; and the preſent Italians are in the ſame practice.

The YELLOW BUNTING, or YELLOW HAMMER.

The tail feathers are blackiſh, the two outer have a white ſpot on their inner web at the tip.

It inhabits Europe, and is very common in England. It builds a large fiat neſt on the ground, in meadows, of hay and ſtraw, mixed with moſs, dried leaves, and ſtalks, very indifferently put together, and lined with hair, or wool. In the winter it frequents villages and barns; in the ſummer it deſtroys the larvæ of the common cabbage butterfly.

The RICE BUNTING.

The plumage is duſky, the back of the neck reddiſh, the ſtomach black, the tail feathers ſharply pointed.

The female is of a very different colour from the male, her plumage is grey, but her tail feathers are like the male's.

D 6 Theſe

These birds inhabits Cuba, and make great havock amongst the rice plantations; for rice is their favourite food, when it is in its milky state. In a very short time they destroy whole acres of rice.

In autumn, after doing a great deal of mischief in Cuba, the females migrate to Carolina; they pass over the sea in immense flocks, and are often heard by the sailors in their passage. Their stay in Carolina is very short, about three weeks, in that space of time they become so fat, that when they are shot, they often burst in the fall.

In the spring, both the male, and the female Rice birds, make another short visit to Carolina; some few stay there the whole year.

This bird was never known Carolina until rice was cultivated there. By a wonderful instinct, they have followed this plan; in the same manner Sparrows have passed into countries, where they were unknown, until they became better inhabited, and corn was cultivated; and Partridges have continued to spread themselves with the cultivation of corn, in countries where before they were utter strangers.

It is not a century since rice was first introduced into Carolina, a small quantity of some that was brought as a sea store, in a vessel, was left in

the

the bottom of a fack, perhaps not more than half a bufhel; from this fmall beginning, the immenfe quantities were produced, which we have for a long time been fupplied with from South Carolina.

The REED-BUNTING, or REED-SPARROW. This bird, as its name implies, frequents marfhy fituations; it makes its neft among the reeds, fufpended from four of them like a hammock, and a few feet above the water. The neft is compofed of the dry ftalks of grafs, and lined with the down of reeds.

The male bird in the fpring, fings perched upon a reed, and by night as well as by day.

It inhabits England, and moft parts of Europe, between Sweden, and Italy; but in many countries it is migratory.

The WHIDAH BUNTING inhabits Angola, and other parts of Africa; by fome it is called the Widow Bird. The two middle feathers in the tail are four inches long, very broad, ending in a long thread; the two next are about thirteen inches in length, broad in the middle, narrower at the ends, and pointed; and from the fhaft of thefe, proceeds another long thread, the

other

other tail feathers are little more than two inches long.

They moult in November; the male then lofes the long tail feathers, until the next moulting time in the fpring: and from November, until June, is little diftinguifhed for the beauty of his plumage, or the fweetnefs of his fong.

Travellers report that the birds of this fpecies make a curious neft of two ftories, the upper for the male, and the lower for the female.

The FAMILIAR BUNTING inhabits Java, it is of a gentle difpofition, and eafily tamed. A gentleman faw one in a cage at Java, that was fo fearlefs as to perch upon the hand.

When any perfon whiftled to it, it fung fweetly in return, and if a difh of water was prefented, it immediatly bathed itfelf. It was fed with rice.

GENUS 73.   TANAGRA.

The beak is conic, a little inclining towards the point, at the bafe a little three fided; pointed, and notched near the point.

The

The birds of th's Genus inhabit in great numbers the warmer climates of America. There is a great refemblance in their habits, and character, to the common Sparrow; like them they feed upon grain, fly low, and take fhort, and interrupted flights; and are of a fociable difpofition, approaching the habitations of men; but they excel them in the beauty of their plumage.

They frequent only dry fituations, and lay but two, or at moft three eggs at a time; but as they live in fo warm a climate, they breed in every feafon of the year; and thus compenfate for the fmall number which they produce at a time, and this is indeed the cafe with many birds, that inhabit the torrid zone.

The RED-BREASTED TANAGER.

Its plumage is moftly black, the forehead, throat and breaft red.

Thefe birds inhabit Cayenne, Guiana, Mexico, and other parts of America; feed on fruits, approach the habitations of men, and are generally feen in pairs.

Their neft is of a cylindrical form, compofed of leaves, and dried fibres; and fufpended from the branch

branch of fome low tree, with an opening at the bottom.

The RED TANAGER.

The general colour of its plumage is a pale red: The wings and tail, are block; the tail feathers are white at the points.

It inhabits Canada, and the woods near the Miffifippi. The fong of this bird is very pleafing. It makes a winter provifion of Maize, which it conceals with great care, covering it with dry leaves, and only fuffering a fmall hole to remain for the entrance. It is fo attentive to its treafure, as never to go very far from it, except to drink.

The BLACK-CROWNED TANAGER feems an exception to the remark of Buffon, that the Tanagers are all of the American birds, fince this is found in Afia, neighbourhood of the Caucaffian mountain, Teflis in Georgia. It builds its neft in a tree, called the Chrift's thor furnifhed with ftrong, and fhar the brood from birds of prey.

The

The JACARINI TANAGER inhabits Braz', frequenting land newly cultivated ; its manner of hopping upwards, from branch to branch, first :lighting on one foot, and then on the other, al-ternately, is very fingular ; every l.p i accompanied with a little note, and at the fame time it fpreads its tail.

There are feveral other Tanagers, but not very diftinguifhable from thofe already mentioned, in their manners, and character.

GENUS 74.   FRINGILLA.

The bill in birds of this Genus is perfectly conic, (not like the Grofbeak's, rounded from the bafe to the point in each mandible, flender towards the end, and fharp pointed.

THE C H A F F I N C H.

The bill is of a pale blue, with a black tip ; the inner and outer webs of the quill feathers, white on their lower parts, forming with other of the wing feathers three white lines, acrofs the wing. The outmoft feather on each fide the tail marked obliquely with white from the top to the bottom ; the next has a white fpot on the inner web, the reft are black.

Chaffinches inhabit all arts of Europe, and fome parts of Africa : In England they are very common,

common, continue all the year, frequenting our
court yards. In Sweden, the females migrate to
the southward, and return again in the spring;
the males stay all the year; and even in Hampshire flocks of females only, have been observed;
at least from their plumage they have been supposed to be so; but when we consider that the seasons occasion an alteration in the colours of some
birds, there may be reason to suspect that this may
be the case with the Chaffinches: Indeed this
seems more probable, than that the females should
separate from the males of the same species, to
migrate into another country, when they might
subsist as well as the males, in those climates
where they were native. Their nest is curiously
constructed of fibres of plants, and moss, neatly
lined with hair, wool, and feathers. They build
in bushy shrubs, or trees well cloathed with foliage, artfully concealing it, and lay five or six
eggs. The male is numbered among our song
birds, though its notes are not very melodious;
it has been remarked that when a Chaffinch has
accidentally lost its sight, it sung more constantly
than before; and this has induced some incon-
siderate persons, to be guilty of the cruelty of
closing their eyes, by burning the eyelids with a
hot wire; this occasions them to inflame, and unite
together;

together; thus to amuse a thoughtful person
with an indifferent song, a poor bird is con-
demned to perpetual blindness.

The beak of the Chaffinch is very strong; they
are lively birds, and in constant action; they are
not often perched, but run upon the ground;
they live mostly upon grain and seeds, and feed
their young with insects and caterpillars.

The BRAMBLING.

They inhabit woods in Europe; at certain seasons
they migrate into England, in vast flocks; and
are occasionally seen with Chaffinches. In
France they often fly together, in immense numbers, and six hundred dozen have been taken in
one night. They make their nests on tall fir
trees, with moss on the outside, lined with wool,
and feathers; their eggs are four or five in
number.

They feed on beech mast, and various seeds.

The GOLDFINCH.

The bill is white, tipp'd with black; the base surrounded
with rich scarlet feathers; the cheeks are white, the top of
the head black; the back is brown. The quill feathers
black, marked in the middle with a rich beautiful yellow;
the

the tip of the quill feathers white. The tail black, but
n oft of the feathers have a white spot near their end.

They inhabit Europe, and are found but in
fmaller numbers, in Afia, and Africa.

This beautiful bird, however, diftinguifhed for
the richnefs of its plumage, interefts us as much
from the agreeablenefs of its fong, and the docility
of its character.

Its flight is low, but generally regular, continued, and in one direction, not up and down, and
by ftarts, like the Sparrow's.

The neft of the Goldfinch is very round, firm,
and neatly executed, much refembling that of the
Chaffinch; it is generally built on a fruit tree,
on a flender branch; fometimes they build in
bufhes, and have generally two broods in a year.
They are very careful of their young, and feed
them with caterpillars, and infects. In a cage
they are fed with hemp-feed, in a ftate of liberty
they eat many kinds of grain, and are particularly fond of the feeds of thiftles, and of
turnips.

The Goldfinch may be made to breed with a
female Canary bird; their offspring refembles the
Goldfinch, with refpect to the head, and wings,
and is capable of producing; but not their de-
ſcendants.

fcendants. The C   ry a lit- b'rd,
and in an aviary alw.)   s to roof on
the higheft perch.

Their docility is very furprifing; th   ave been
taught to dance, to counterfe t death, to fire a
little cannon, to move a fmall litter, to open a
box in which their food is kept, and to draw up
a little bucket of water out of a glafs fufpended
at fome diftance below their cage.

Their fong is pleafing; they begin to fing in
the month of March, and continue all the fummer.   Bird-fanciers pretend that the Goldfinches
in Kent, excel thofe of every other county.

They are fubject in confinement to epileptic
fits.

The   CANARY-BIRD.

The beak is whitifh, the body a yellow inclining to white.
The wing and tail feathers are of a green caft.

It inhabits the Canary iflands, feeding principally upon the feeds of the phalaris, or canary
feed; in confinement it will eat hemp, flax, and
rape feed; it breeds with the Goldfinch, and their
young are fertile, but not in the fecond generation.

Of this very interefting bird there are many
varieties; the natural confequence of its being
domefticated',

domesticated, and bred in confinement. Some
are grey, upon a ground of yellowish white, in
others the plumage is of a brilliant yellow, in others
again of a mealy colour. Some are distinguished
by a crest, in others the eyes are red; amongst
these a number of other combinations take place,
so that near thirty varieties have been described.

Great numbers of these birds are imported every year from Germany, particularly
from Tirol, and though the Germans who deal
in them carry them a thousand miles upon their
backs, yet they are sold as low as five shillings
each. At Infpruck they are a little article of
commerce, from that town they are fent to feveral parts of Europe, and in confiderable numbers to Conftantinople. They are bred very generally in England, for this purpofe they fhould
be provided with a cheerful chamber, furnifhed
with boxes, or little bufhes formed of twigs of
birch, or heath, to build their nefts in; and little
nets filled with fine hay, mofs, feathers, down,
flax, and goat's hair, fhould be hung in different
parts of the room ; it is very proper to have fewer
males than females, that the male birds may not
difturb the females, which fometimes they are
apt to do, and occafionally to break their eggs.

They

2.

They should be fed at this time with turnip feed, oats, millet, and bean-meal.—The day before the female is expected to hatch, she should be supplied with scalded food, and eggs boiled hard; but all green vegetables must be avoided, because they would weaken the young. Sometimes a little bread moistened with water will be a very proper food.

When they are separated from their young, they ought to be supplied with plantain, and lettuce feed, in small quantities.

In the management of Canary-birds, it is possible to do too much; we should never lose sight of nature; in their wild state they frequent streams, we should therefore let them have a constant supply of water, that they may occasionally bathe themselves.

There are several disorders to which they are subject; sometimes they die in consequence of eating to excess, sometimes in moulting: young Canary-birds are liable to a complaint in their intestines, which become inflamed, and distended: this is a common, and generally a fatal complaint, arising from too great a quantity of food, or from its being too moist. Some indispositions to which they are subject, proceed from want of neatness. The pip is another complaint; the bird

afflicted

afflicted with it is dull, the feathers are ruffled, and the body appears enlarged, the head drawn in, or concealed beneath the wing. When these symptoms appear, it is proper to examine if there be not a little pimple near the tail; this often breaks of itself; sometimes the bird presses it with its bill, and by breaking it, effects a cure.

Like many other birds in a domestic state, it is attacked by epileptic fits. Such are the unhappy effects of confinement.

The cares and attentions which we bestow on these pretty warblers, are amply repaid by their song. They are the domestic Nightingales, filling our houses with their delightful melody.

The Canary-bird has a very nice ear, a strong memory, and great powers of imitation; it may be taught the song of the Nightingale, or the Lark, or to pipe a tune like a Bullfinch.

Its manners, and character, are extremely interesting; it contributes to the amusement of the young, it enlivens solitude, and recals to us in our chambers the pleasing impressions of the melody of the groves. The Canary-bird will live in confinement if properly attended to, from ten, to eighteen years.

A female grey Canary-bird that escaped from its cage, paired with a Common Sparrow, and
                                        produced

produced a brood in a bird-pot which was placed against a house.

Near Paddington, in Bedfordshire, some Canary birds have been seen in hedges, in a wild state; they had probably escaped from an aviary. The English Canary-birds have mostly the Wood-lark's song, those from Tirol that of the Nightingale.

THE SISKIN.

On the middle of the wing feathers is a bird of an olive colour, the tail is a little forked, the two middle feathers inclining to black and edged with olive colour, the rest are yellow tipp'd with black, and edg'd with green.

The throat is dusky in the male, in the female white.

The Siskin inhabits Europe, frequenting those parts that abound with juniper. It visits these islands at uncertain times, but generally in barley seed time; from whence it is called in Sussex the Barley bird. It is sold in the bird shops in London, under the name of the Aberdavine, but has no very pleasing song. The nests of this species are concealed with great art, for though there are many of the young birds in the woods, on the banks of the Danube, that seem but just able to fly, their nests are seldom to be found. Siskins run along on the under part of the branches of trees, like the Tit-moufe fpecies, with their backs downwards.

The Sifkin will breed with the Canary-bird.

The RED-HEADED LINNET.

The breaft is tinged with a fine rofe colour, is has a bar of white on the wings. The quill, and tail feathers, are black, edged with white.

They inhabit moft parts of Europe, breed in the north of England, and are often feen in flocks on the fea-coaft.

This Linnet eafily pairs with the Canary; there are few birds more deferving our attention from the beauty of their plumage, the melody of their fong, or the agreeablenefs of their character. A gentleman pair'd a male bird of this fpecies, with a female Canary bird, which had been accuftomed to be releafed every day from its cage, but to return regularly for its food, and to rooft. She made her neft in a neighbouring bufh, and brought her young ones with her to the window to feed. They had the mealy plumage of the Canary-bird, but the rofe-coloured breaft of the Linnet.   -

In a domeftic ftate, this Linnet lofes the beautiful colours of its head and breaft; it is very capable of imitation, and there was one at Ken-fington who having been taken in the neft, and

brought

brought up where there was a little child, learnt to articulate the words pretty boy, from having heard them frequently addreſſed to the infant by its nurſe.

Linnets frequently build in vines, in gooſeberry buſhes, and nut-trees; the neſt is compoſed of roots, leaves, and moſs on the outſide, and lined with wool; they breed in May, July, and Auguſt; about the latter end of that month, they collect together in companies, and live ſo during the winter. They fly very cloſe to one another; riſing, and alighting together, and perching upon the ſame trees. In the ſpring, they generally rooſt on oaks, or trees which are not deſpoiled of their withered leaves, often ſinging in concert; they feed on flax, which their name implies, on hempſeed, thiſtles, and many other ſeeds; but in confinement hempſeed is too nouriſhing a food, it fattens them to ſuch a degree, as to prevent their ſinging, and to deſtroy their health.

The Linnet frequently ſhakes the duſt amongſt its feathers, and ſhould be ſupplied in confinement with ſand for that purpoſe, and water to bathe in. With a pipe, or bird organ, it may be taught to whiſtle tunes; it is very capable of attachment, and ſeems by its animation to ſhew its affection to thoſe who ſupply it with food.

E 2　　　　　　The

The LESSER RED-HEADED LINNET much resembles the last species.

It is gregarious, the nest has been found on an alder stump, near a brook, about two or three feet from the ground; it was constructed on the outside with dried stalks of grass, and other plants, intermixed with wool; the lining was hair, and feathers. The bird was sitting on four eggs, and so attached to her nest, as to suffer herself to be taken in the hand, when released she did not forsake it.

Their plumage is not so beautiful as that of the last species.

They inhabit most parts of Europe, but in some they are migratory; they appear in Germany in October, and leave it in February. Like the Titmouse, they run along the under parts of the branches of trees.

They become very fat, and are a delicate food.

The RING SPARROW

Is grey, with white eye-brows, the throat clay colour'd.

It inhabits Europe, principally Germany, and Italy.

It

COMMON SPARROW.

It feeds on feeds, and infect ; but feens a tender bird, being often found dead in fevere winters, in the hollows of trees.

The CUBA FINCH.

The plumage is of a purplish caft, acrofs the breaft it has a purplifh chefnut colour'd band.

Thefe birds inhabit the ifland of Cuba, and uniting in confiderable flocks, are very de ru hve to rice plantations: indeed they are com on to both the continents; and found in the E.ft as well as the Weft Indies.

The COMMON SPARROW.

The wing and tail feathers are dufky, the body is grey and black, the checks are white; under each eye is a black fpot, and a bind of white on the wings. The male is diftinguifhed by the black on his throat.

The Common Sparrow is fo well known, that a particular defcription of his colours would be unneceffary; though in this fpecies there are fome varieties, the effect perhaps of accident; for fome have been found white, fome varied with brown, and black, fome black, and others yellow.

E 3   In

In Lorraine Black Sparrows are frequently found, but the colour perhaps may be only external, and the effect of smoke, since there are many glass-houses there, which they frequent, hence their feathers may be discoloured.

The sparrow is never found in any places that are far from the habitations of men; they do not frequent woods, or large extended plains; but like rats, establish themselves about our houses, in order to subsist at our expence. They infest the corn in the Orknies by thousands, but were unknown in Siberia, until the Russians attracted them by the cultivation of corn.

As they are very indolent, and very greedy, it is upon the labours of others that they wish to subsist; they frequent our barns, our corn ricks, our poultry yards, and dove houses; they follow the labourer when he sows, the mowers during the harvest, the threshers in the barn, and observe the poultry when they are fed. They even pierce the crops of young Pigeons to take out the grain, and prey upon bees, an insect peculiarly useful to us. Of all birds they are the most mischievous, and yet of all the most difficult to destroy, or to remove. Nothing will dislodge them from the places they frequent. They are cunning, timid, difficult to be deceived; they easily detect traps,

traps, and give great trouble to thofe who attempt
to catch them, and even then almoft every effort
is in vain, for they breed thrce times a year, and
if you deftroy their neft, in two days they will
make another; if you break their eggs, in ten
days, they will lay again; if you fhoot at them
when they are upon the roof of the houfe, they
only conceal themfelves the better in your
barns.

Thofe who have kept Sparrows in cages, fay
that a pair will confume near twenty pounds
weight of corn, in a year; we may judge from
hence how deftructive they muft be. It is true
they feed their young very much on infects, and
that they themfelves deftroy many infects, yet
grain feems to be their favourite food. On account of the mifchief they do, a price is fometimes fet upon their lives, in many villages; in
fome parts of Germany, every year, each peafant
is obliged to produce the heads of a certain number of Sparrows.

Yet however mifchievous they may feem, as
Providence has made no animal whatever but to
produce fome good, we may be affured that the
Sparrow anfwers fome defirable purpofes.

Mr. De Buffon attempted to deftroy them by
burning fulphur, mixed with fome rofin and char-

coal

coal under the trees where they roosted, but in vain; they only at first perched higher in the trees, and afterwards flew upon a roof at some distance.

Sparrows are easily tamed, and brought up in a cage, they may even be taught when young something of the song of other birds; though they are easily made familiar, yet they do not naturally associate together; they are generally alone, or in pairs, though they sometimes plunder in companies.

They generally build under the tiles, in the roofs of houses, in thatch, in holes of walls, under the eaves, in the corners of windows, and sometimes even in trees.

When they build in trees they make a little covering to their nest, to keep out the rain; some again too indolent to make nests of their own, build in the old nests of rooks, and drive the pigeons from their boxes. They often take possession of a Martin's nest. It is said that the Martin, sometimes resenting the injury, brings clay, to stop up the entrance, and so bury the Sparrow alive.

The TREE SPARROW,

The quill and tail feathers are dufky, the body is grey and black, there are two white band on their wings, and a large black fpot near each ear.

It inhabits moft parts of Europe, and is frequent in Lancafhire, Lincolnfhire, and Yorkfhire.

It lives in the country, frequents highways, perches upon fhrubs, and low bufhes, and builds its neft on trees.

Tree Sparrows prefer an open country, they are lefs numerous than Sparrows, and probably do not breed fo often; they affemble in flocks at the clofe of fummer, and remain together during the winter.

This bird is almoft always in motion, when he is perched he is continually moving, turning, raifing, and dropping his tail.

It is much lefs mifchievous than the Sparrow, not deftroying much corn, but preferring fruit, and wild feeds. It perches on thiftles, and eats their feeds, and often preys on infects. The fong is nothing extraordinary. It lays out four or five eggs.

The

The FRIZZLED SPARROW.

The plumage is of an olive colour. The under part of the body yellowish, its head is black and many of its feathers frizzled.

It is suppofed to inhabit Angola, or Brazil.

There are many other fpecies of the Genus Fringilla, but thofe mentioned are the moft remarkable for their plumage, and their manners.

Genus 75.   MUSCICAPA.

The bill of the Fly Catcher Genus is flatted at the bafe, almoft triangular, notched at the end of the upper mandible, which bends towards the point; and befet with briftles.   .
The noftrils roundifh.

This is a Genus related a little to carnivorous birds, apparently more innocent, and perhaps quite as ufeful. They are in fact birds of prey, though they abftain from flefh, and are not injurious to fruit, or grain, living upon flies, and gnats, and other infects.

Some

ORDER 3. *Passeres*
Genus 5. *Muscicapa*.

Some of them approach in fize to the Butcher bird, others are fmaller than the Nightingale.

There is no melody in their notes, their character is favage, and folitary. They are feldom on the ground, as their fubfiftance is in the air, they are moftly feen upon trees, which they have little inducement to quit.

In cold countries there are but few fpecies; as their food confifts of infects, it is natural to fuppofe that they fhould exift in the greateft numbers in warm climates; in Europe the fpecies are few, more numerous in Afia, and in America, they are confiderably multiplied.

THE PARADISE FLY-CATCHER.

The head and creft are black, the body white the tail long and wedge under like the tail of a pheafant, the middle feathers are very long.

It is found at Senegal, the Cape of Good Hope, and at Madagafcar. It frequents the mangrove trees on the folitary parts of the banks of the rivers Senegal, and Gambia, feeding upon infects.

THE

The FORK-TAILED FLY-CATCHER,

Has a long forked tail, the plumage of its back black, beneath white.

It inhabits Canada, and Surinam, and is about the fize of the crefted Lark ; it is feen continually flying down from trees in the neighbourhood of plains that are covered with water, alighting up-on the little hillocks, or the plants that fwim on the furface, and flirting its tail continually, like the Water Wagtail.

The CRESTED FLY-CATCHER.

The neck is blue, beneath yellow, the back greenifh, the wing and tail feathers a reddifh brown.

It breeds in Carolina, and Virginia, retiring in winter to a warmer climate ; its cry is very difagreeable ; it affociates with no other bird, and feems of a favage, and fullen character.

Its neft is made of fnakes fkins, and hair, in holes of trees.

The

The RED-EYED FLY-CATCHER.

The plumage is olive colour, more light coloured beneath. The eyebrows are white, and the eyes red.

It inhabits Carolina, but migrates into Jamaica at the approach of winter, where from its singular note it is called Whip Tom Kelly. It generally builds in apple trees, suspending its nest from between the fork of some bough, amongst the leaves. The nest is pendulous, and curiously formed of cotton and wool, lined with hair, and dead grafs, and ingeniously bound to the branches by a mofs-like thread, twifted round them, and about the outfide of the nest.

The CAT FLY-CATCHER.

Its plumage is dufky above, ash coloured beneath, the head black, the under part of the tail dirty red.

This bird inhabits New-York and Carolina; frequenting bufhes and thickets; its note refembles the mewing of a kitten, from this it takes its name. It has great courage, and will affail a crow, or almoft any bird, however larger than itfelf.

Infects are its favourite food. The outfide of its nest is made with leaves, and matting ruthes, the infide with fibres of roots.

THE

The SPOTTED FLY-CATCHER.

The plumage on the back is of a moufe colour; beneath whiteifh; the neck is freaked; the feathers under the tail are of a reddifh caft.

This bird is migratory; it arrives in April, and leaves us in September; it frequents woods, feeking folitary, and fheltered feenes, though fometimes it infefts our orchards, and is very deftructive to cherries; on this account in Kent, it is known by the name of the Cherry Sucker.

It feems a melancholy, and rather a ftupid bird; building its neft without any appearance of concealment, againft trees, or upon bufhes. Their nefts are clumfily made, and not always with the fame materials, fometimes with mofs only, and fometimes with wool, and large fibres of roots intermixed.

They chiefly feed on infects, and collect them on the wing. They feldom alight upon the ground, or ufe their legs in running; but perched upon a branch, or poft, watch the infects as they are flying near, fpring after them, and again return to their place.

The FAN-TAIL FLY-CATCHER,

Inhabits New Zealand; it is a very familiar bird, and is eafily tamed; it will then fit upon

any

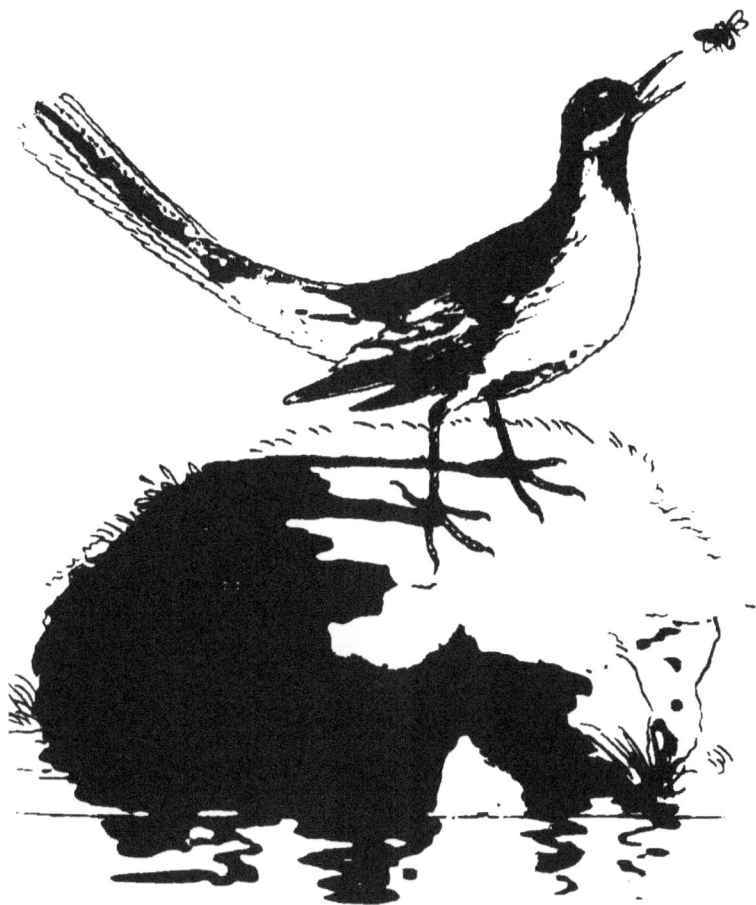

WATER WAGTAIL.

Published June 1st 1797 by John Johnson St Pauls Church Yard London

any perfon's fhoulder, and pick off the flies as they alight, or purfue them as they are paffing by, fpreading its tail as it flies, like a fan.

The PURPLE THROATED FLY-CATCHER,

Inhabits Cayenne, and other parts of America; collecting together in flocks, they generally precede the Toucan; they feed on fruit, and infects; they are lively birds, in almoft conftant action, frequenting the woods, and frequently repeating a fharp cry, like the word Pihauhau.

The TYRANT FLY-CATCHER,

Is claffed by Linnæus among the Butcher birds. It has a black bill and head, the crown divided lengthways by a ftripe of fcarlet, orange, or yellow; the back is afh colour, the tail black tipt with white.

The Tyrant builds its neft of mofs and roots, without concealment, on fhrubs and bufhes, fometimes on the faffafras; it appears in Carolina and Virginia, in April; breeds there, and retires at the beginning of winter. Its courage is furprifing; it attacks, and chaces away every bird of whatever fize, that approaches the fpot which it inhabits. None efcape his fury, none attempt refiftance, when he affails them on the wing; for

it

it is only whilst they are flying, that he makes his attack.

Mr. Catesby saw one fix itself on an Eagle, and persecute him so, that the Eagle turned upon his back, and put himself into a variety of postures as he flew, but in vain; at length he alighted upon a tree, and kept perched there until the little tyrant left him through fatigue.

Whilst the female sits, the male perches upon a bush, or shrub, near the nest, and chaces away any little bird that approaches; but if a large bird, an Eagle, Hawk, or Crow, come within a quarter of a mile, immediately he makes his assault, and drives it to a distance. When his young are flown, his impetuosity ceases, and he becomes as peaceable as other birds.

THE CHATTERING FLY-CATCHER,

Inhabits the interior parts of Carolina, two or three hundred miles from the sea; it lives by the banks of great rivers, and screams so loud that the noise is reverberated from rock to rock. It is so shy, that the fowler with difficulty can approach it; it flies with its legs hanging down, and often rises perpendicularly, and alights by jerks.

There

There are many other species, but little distinguished by their habits, from these mentioned, or remarkable for their plumage.

These birds, as well as Swallows, Nightingales, &c. which feed on insects, are of essential service to man.

Without their assistance, we should be assailed by myriads of insects; we should be tormented by their bites and stings; they would devour the produce of our lands, defile our provisions, which in vain we should endeavour to preserve; they would infest our chambers, and disturb our rest; for so rapid is their multiplication, that our attempts to destroy them, could scarcely be successful. To this useful tribe of birds, we are indebted for our deliverance; by their means insects are kept in such a degree of subjection, as only to consume that which is superfluous, and which without their help would become offensive.

The advantages we have derived from our little deliverers during the summer, we feel in autumn; towards the close of that season, they leave us to fly to milder climates, in pursuit of the insect food, which they seem to have a pre-sentiment is soon to fail them here.

During the interval between their departure, and the first frosts, insects increase in abundance, myriads of gnats infest the air, and their rapid multiplimultip'ication would be feverely felt, did not the friendly frofts of winter, deliver us again from their devaftation.

GENUS 75.   MOTACILLA.

Pennant has divided the Genus Motacilla, (and with reafon,) into two Genera; the one he calls the Wagtail, the other the Warbler. We fhall obferve his example, fo far, as to arrange this Genus under two divifions.

DIVISION THE FIRST. THE WAGTAIL.

The birds of this divifio have a weak flender bill, flightly notched at the tip.

The tongue is jagged at the end.

The legs are flender.

The tails are long, and frequently in motion.

'l hey frequent the fides of brooks, and are almoft conftantly running; they feldom fly; but when they do fly to any diftance, it is in an undulatory, or wavy direction; they rarely perch, their nefts are on the ground, and they make a twittering noife as they fly.

THE WHITE WAGTAIL.

The bill, head, neck, chin and breast are black. The forehead, cheeks, and sides of the neck are white, the two outer tail feathers obliquely marked with white.

It inhabits Europe, is migratory, appears early in the spring, frequenting villages, and constantly flirting its tail.

The Swedish farmers say, when you see the Wagtail return, you may put your sheep into the fields; when you see the wheat-ear, you may sow your grain. So useful is the information we may acquire, by observing the animal creation. In England the Wagtail continues all the year, but in the winter it moves from the North to the South.

The Wagtail is a very small bird, but his long tail makes him appear much larger than he really is. As he flies he spreads his tail, which is a considerable support, and enables him to turn, to gambol, to rise and fall in the air.

Wagtails run lightly along the ground, moving very nimbly, close to the edge of the water; and sometimes they just step into the little wave, formed by the water as it spreads on the shore; but they most frequently flutter upon the dams of mill pools, or frequent the sides of shallow streams,

hopping

hopping from ftone to ftone; approaching fami-
liarly the women who may be employed there in
wafhing, fporting around them, and picking up
the crumbs which they may let fall.

They make their nefts upon the ground, under
roots; or more frequently by the water fide, near
piles, that have been driven down to keep up the
banks.

The neft is compofed of dry grafs, fmall roots
intermixed with mofs, negligently made, and lined
either with feathers, or horfe-hair. The female
lays four or five white eggs, fpotted with brown;
and has feldom more than one brood in a year.
Upon the approach of danger, both male and female flutter before the enemy, as though they intended to miflead; but if their young fhould be
taken, they follow the fpoiler, flying above his
head, in circles, and calling to their young in
the moft melancholy tones.

When their young are hatched, they attend
them with the greateft care.

They keep the neft perfectly neat, removing
every thing that would be dirty, and difagreeable:
a gentleman placed fome little pieces of white
paper near their nefts, which they carried away
to a diftance, as though it were difpleafing to
them.

When

When the young are able to fly, their parents still attend them for three weeks or a month, feeding them plentifully with insects, and ants eggs. They seem always to eat in great haste, scarcely giving themselves time to swallow the worms, flies, and gnats, which they catch. They are constantly rising, and wheeling in the air, to take insects; when they fly it is by jerk, up and down; whilst they are flying we often hear their note, but more frequently perhaps just as they have narrowly escaped a bird of prey.

They are not much afraid of man, and even after they have been shot at, they fly but a little way, and return again; they may like the lark be taken by a net, and looking glass.

In France they are migratory birds. In autumn they assemble together among willows, by the sides of the water, or on the roofs of mills, or other situations near the water; they are then remarkably sprightly and sportive, twittering to, and pursuing one another; at length by common consent, they all fly away together, to warmer climates; and in the winter it is said they are found in Egypt, and at Senegal, which are both in Africa.

They constantly attend the plough, to pick up the worms.

At

At Barr in Staffordshire, a Water Wagtail that frequented a house, by being occasionally fed with crumbs and pieces of meat cut very small, grew tame, and continued there the whole of the winter, for several years ; he frequently flew upon the window cills to search for dead flies, which sometimes fell from the joints and crevices upon opening the window.

THE YELLOW WAGTAIL.

The breast and stomach are yellow, the two outer tail feathers obliquely streaked with white, the throat is spotted with black in the male, the crown and upper part of the body are of an olive green.

In the winter, the yellow Wagtail, when the common Wagtail is gone, approaches villages, and seeks its food by streams of water, that are not frozen, and shelters itself under the banks of rivulets : In that dreary season, if the cold be not extreme, its gentle warbling is heard ; it is in a low key like the autumnal song of the common Wagtail, and very different from the shrill note which it utters when it rises on its wings. In the spring the yellow Wagtail makes its nest in meadows, and sometimes in copses, at the root of a tree, near a rivulet, and in cornfields. The nest is placed upon the ground, and built with dry grass, or moss,

and

and ftems of plants and fibrous roots on the outfide, lined with feathers, or horfehair within; and
more neatly conftructed than that of the white
Wagtail. They lay feven or eight eggs, of a
dirty white, fpotted with yellow. After the hay
harveft, the parent leads them amongft flocks of
fheep. Then they live on flies, and gnats. Whilft
they frequent the water fide, they feed on worms;
fometimes they fwallow feeds, and even beetles
have been found in their ftomach.

The tail of this bird is longer than its body; and
like the white Wagtail it frequents rivulets with
pebbly beds, perching upon the ftones; though
fome ftay all winter, yet many feem to migrate, for
they are not feen in fuch numbers then, as in
the autumn amongft flocks of fheep.

It migrates in the north of England, but continues in Hampfhire all the year.

This laft fpecies is called in France La Ber-
geronette, from their habit of frequenting flocks
of fheep; they follow them in the fields, mix
with them as they graze, and perch occafionally
on their backs. They feem fearlefs of, and familiar with the fhepherd, flying before him, and
in fome countries act the part of a fentinel, by
apprifing him of the approach of the wolf or of
birds of prey.

Yet

Yet though fearlefs of man, they are incapable
of confinement in a cage: deprived of liberty
they pine, and die. Sometimes they have been
known to fly on board a fhip at fea, to familiarife
themfelves with the failors, and only quit the fh'p
when it reached its port.

SECOND DIVISION OF THE
MOTACILLA, OR THE WARBLER,

Their bills are flender and weak.

The noftrils fmall, and funk.

The outer toe united beneath the laft joint, to the middle
toe.

The Warblers are very different from the
Wagtails, and very properly placed under a different divifion.

For the moft part they perch on trees; proceed by leaps, and not by running, feldom twittering as they fly.

They are very numerous, and inhabit the
warmer regions, where infects abound; infects
being their principal food.

THE

The NIGHTINGALE,

Its plumage is of a reddifh afh colour, with bands of afh colour on the legs. It inhabits thofe parts of Europe that are fhaded with woods, finging in the evenings of the fpring, with amazing powers, for fo fmall a body ; and falling fome-times a victim to emulation, in fong.

The Nightingale is a bird not very generally dif-perfed ; there are none in Africa, and in feveral parts of Europe, and even of England, they never appear. Some writers affert, that there are none in Devonfhire, or Cornwal, or to the north of the Trent, except near Doneafter in Yorkfhire, nor in fome parts of Holland, Ireland, Scotland, or the northern parts of Wales ; though they vifit Sweden.

In England, France, Italy, Germany, and Greece, they are only birds of paffage ; it is fup-pofed they retire into Afia, as they are found there in Perfia, China, and even in Japan, where they are much efteemed for their finging, and fold for a confiderable price.

Indeed they are every where migratory birds ; and this difpofition feems fo implanted in them, that thofe who are confined in cages, are very much agitated, and reftlefs, at the ufual periods of their migrations. This inftinct feems to act in them, at least independent of cold, or want of nourishment, since in a cage, the temperature must be always nearly the same, and their food generally in equal abundance.

The Nightingale is not originally an American bird; there are some in the northern parts, these may have migrated from the north eastern parts of Asia, where a very narrow sea divides the Continents. The song of the Nightingale in Canada, is said to be much inferior to the song of the Nightingale in Asia, or Europe, and their song in Scotland, to their song in Italy; indeed a scarcity of food, and a cold climate, materially affect the singing of birds.

As the male Nightingale passes much of the night in singing, it has been supposed that they never sleep in the summer season; and with equal folly, that the heart and the eyes of a nightingale placed under a pillow, would prevent the person from sleeping who lay upon it: it is right to mention these follies, in order to expose and contradict them. Attentive observers find that in the summer season, Nightingales sleep occasionally in the day, and in confinement they have been observed to dream, and to warble in their sleep in a low voice.

When

When in a cage they bathe themselves after they have sung; they have been observed to do this immediately as candles were lighted. It may be well to mention another effect of light upon them; a Nightingale who sung very well having escaped from his cage, immediatly flew into the fire, and was burnt.

Nightingales, at least the males, have a habit of raising, and sinking their bodies alternately, ballancing themselves on their perches, and both males and females raise and let fall their tails.

Nightingales conceal themselves in the thick set bushes, they frequent hedges, and low coppices, they feed on aquatic and other insects, on worms, on the eggs or larva of ants, some fruits and berries. They are not suspicious, but so much the contrary, that they are taken in a variety of ways: they may be attracted by the song of birds, by a fine voice, and even by dif-agreeable noises, for they seem foolishly inquisitive; they wonder at every thing that is new, and are often the dupes of their curiosity; they are taken with birdlime, with traps, snares of various kinds, and springes. Sometimes they abound in particular places. Once in a dry season, in the neighbourhood of the Forest of Ardennes, which is in the Netherlands, the shepherd's boys have

each

each caught near twenty in a morning : they re-
forted to thofe parts of the foreft where the
water was not dried up.

Thofe who take them are very careful not to
injure their feathers, left it fhould delay their
fong ; for whilft they are moulting, they are always filent. Sometimes to haften the time of
their finging, the feathers are taken out of their
tail, that the new ones may grow again the fooner.

Nightingales are faid to be very nice food,
when they are fat.

In Gafcony they are fometimes fattened for the
table ; fo wanton a deftruction of a bird whofe
fong is fo melodious, recals the depraved, the
cruel, and extravagant fancy of Heliogabulus, a
Roman Emperor, who feafted on the tongues of
Flamingoes, Peacocks, and Nightingales.

The plumage of the Nightingale is not beautiful, the upper part of the body is of a reddifh
brown ; the throat, breaft, and ftomach of a
whitifh grey. Nightingales that are bred in the
fouthern countries, have a darker plumage ; and
thofe bred in northern countries a lighter plumage.

The tongue of the Nightingale is very remarkable, the end is furnifhed with threads, and
feems as though it had been cut off.

But

But the Nightingale's song gives it a superiority over every other bird; the variety, and the melody of its powers, are equally astonishing. So small a bird that does not weigh more than half an ounce, can fill the compass of a mile with its song.

In the fine nights in the spring, when the weather is serene, and all nature seems as it were hushed in silence, this inimitable songster fills the grove with a melody, which seems to unite the excellencies of all other singing birds, and in effect far exceeds them: he begins with a slow and gentle warbling, at first low, and half pronounced, as though he were trying his powers; by degrees his notes rise, he becomes more and more animated, and loud, and displays such a combination of melody and powers, his notes are so various, so voluble, so soft, his tones sometimes so plaintive, gently dying away, at others so full, so animated, and expressive, and the whole so varied, and wonderfully combined, that it is impossible to conceive sound more melodious, or more interesting. The pauses in his song produce a wonderful effect, they give us time to enjoy those sounds whose impressions are still left upon the ear; we soon wish the song to be repeated; and soon we hear another combination of

F 3 melody,

melody, are varied, and though different, as pleafing as the former.

One reafon perhaps that the fong of the Nightingale produces fo wonderful an effect, may be, that it fings in the night, when its warbling is uninterrupted by other founds.

Nightingales begin to fing early in April, and feldom fing after June; their fong is fufpended by the attentions they are obliged to give to their young.

In confinement the Nightingale fings nine months in a year; and the fong is ftill more perfect than that of Nightingales in their natural ftate. They will fometimes fing in a few hours after they are taken, generally in feven or eight days; not that they are regardlefs of the lofs of their liberty, for at firft they are fullen, and refufe to eat, and would die of hunger if food were not put into their beaks, or kill themfelves by ftriking againft the wires of their cage, if their wings were not tied. Soon, however, they become more tamed, they delight in finging, are charmed with mufical inftruments, and a fine voice. Their emulation is raifed by other birds; and it is faid that by their exertions they have fallen down dead in their contefts.

As

All Nightingales do not sing equ l y well, there are many that connoisseurs will not keep; they imagine too that the Nightingales in some countries, and even in some provinces, sing better than in others. .

In England a Nightingale from Surry, is supposed to be superior to a Nightingale of Middlesex, which is the next county.

Indeed it is asserted, that there is as great a difference in the song of birds of different counties, as in different dialects.

It is possible that a Nightingale may have improved his song, by hearing other birds; in that case he will teach his song to his young; for every Nightingale is music master to his little ones; and such accidents, in many generations, may produce considerable effects.

Female Nightingales have been known to sing, but their song is inferior to that of the male.

An attempt has been made to write the notes of a Nightingale, and afterward to play them on an instrument, but without effect. Their song has been much better imitated by the human organs. Some time since, a man in London succeeded in his imitation so well, as to make Nightingales perch upon him, and suffer themselves to be taken by his hand.

F 4

In

In fome inftances thefe imitations feemed to require great effort.

The Nightingale is frequently caught, and tamed, for the fake of his fong : but great attentions are **requifite** to fucceed ; his cage fhould be painted green, and fhaded with boughs ; mofs fhou'd be placed under his feet, and he fhould be plentifully fed with a kind of food that he l'kes ; in fhort he muft almoft be deceived with refpect to his confinement : in this cafe the tame Nightingale will fing almoft conftantly, except during the moulting feafon, and his fong will be fuperior to that of the wild Nightingale, for he will embellifh it with the excellencies of the fongs of other birds. They may be taught, to fing in parts, and perhaps even to fpeak, but many of thefe accounts are much exaggerated ; fuch for inftance is the ftory that is told of fome canary birds belonging to the fons of the Emperor Claudius, which it is faid fpoke Greek and Latin, and every day learnt new phrafes, and fome not very fhort, to amufe their mafters.

There is a very wonderful ftory of two Nightingales mentioned by Geffner, which it is much more amufing to read, than eafy to believe ; he quotes a letter, written as he profeffed to fuppofe, by a gentleman of veracity ; which relates
that

that one night as the gentleman lay awake at an inn in Ratisbon, he heard two Nightingales in cages, conversing together about the politics of Europe, that they repeated the conversation at night which they heard in the day, and kept up a long and interesting dialogue, on a variety of subjects.

Nightingales in time, attach themselves to those who have the care of them ; one that was given to a gentleman, no longer seeing the lady that was used to feed it, grew sullen, refused to eat, and was soon reduced to that state of weakness, that he could no longer support himself on his perch ; but having been restored to his former mistress, his animation revived, he eat, drank, returned to his perch, and was well in twenty-four hours.

They have been known to refuse the liberty that has been offered them, and when they have been turned out of their cage, in the woods, they have returned again.

When once they are attached, they distinguish the step of the persons used to feed them, flutter at their approach, and even when moulting make some attempts to sing, and supply that de-fect by every expression of joy

In confinement when young, they are fed with a compofition of fheep's heart, crumb of bread, hempfeed, and parfley, perfectly mixed together, and chopped very fmall; it muft be made frefh every day; afterwards egg boiled hard, and mixed with crumb of bread, and parfley, the whole minced fmall.

Another compofition which fhould be their general food, when they are full grown, is made of two pounds of lean beef, half a pound of grey peas, half a pound of yellow or purled millet, fome white poppy feed, and fweet almonds; a pound of honey, two ounces of flower, twelve yolks of eggs, two or three ounces of butter, and a dram and a half of faffron, the whole dried by a fire, reduced into a fine powder, and paffed through a fieve; this will keep a long time.

With thefe compofitions, a Nightingale has been kept feventeen years in a cage; at feven years of age he began to grow grey, at fifteen, his wing and tail feathers were white; his knees were much enlarged; he had knots on his toes like gouty perfons, and it was neceffary occafionally to file the point of his upper mandible; but he feemed to feel no other inconveniences from old age; he was always cheerful, and careffing.

It is faid, that warm drugs, and perfumes, excite the Nightingale to fing ; that meat, and garden worms are good for them, when they are too fat; and figs when they are too lean ; that spiders are phyfic to them, and very proper in April, and that half a dozen are a dofe. When they have fed on any thing that difagrees with them, they bring it up like birds of prey, in little round balls.

Nightingales are folitary birds, they not do migrate in flocks, but arrive fingly in England, in April, or May, and return in that manner about Auguft or September.

They pair foon after their arrival, and then they fuffer no other of their fpecies within a certain diftance ; and it is fuppofed that this diftance is determined by the greater, or lefs plenty of food, and that where their proper food abounds, their nefts are nearer each other.

They begin to build their neft about the beginning of May ; it is compofed of leaves, rufhes, blades of grafs, very coarfe on the outfide, with fmall roots, fibres, horfe-hair, and a kind of down in the infide ; it is generally expofed towards the eaft, and built on the lower branches of fhrubs, fuch as goofeberry trees, white thorn, floes, &c. or on a tuft of grafs, and fometimes on the ground, fo that the Nightingale herfelf, and the young fometimes, become the prey of dogs, foxes, weafels, polecats, and fnakes.

The female lays five eggs, fits very clofe, and only leaves her neft in the evening, when preffed by hunger; whilft fhe is abfent, the cock feems to watch the neft. In eighteen or twenty days the young are hatched. The number of males is faid to exceed the number of females, fo much, that if by accident the cock fhould be killed, the female would foon be fupplied with another, fo that the young birds would not fuffer.

The female, like the female Canary Bird, feeds her young with food that fhe brings up out of her own ftomach; the father affifts in the care of the young, and then it is that he feldom fings: probably that he may not difcover the neft; though if it fhould be approached, he does not practice any of that artifice which fome other birds employ to miflead, but oftentimes betrays it by his fears, and his cries.

In fifteen days the young are fledged.

In Auguft, the old and the young Nightingales leave the woods, and frequent bufhes, hedges, and fields lately ploughed, perhaps in order to procure a greater plenty of worms, and infects.

If an attempt be made to rear young Nightingales, it will be beft to take thofe of the firft hatch,

and

and their beſt inſtructor will be an old Nightingale.

The HANG NEST.

Upper part of the body a duſky green, beneath duſky orange, a black line above and beneath the eye.

It inhabits America, and makes a curious pendulous neſt.

The HEDGE SPARROW.

Above a duſky grey, the coverts of the wings white at the points: The breaſt a blueiſh aſh colour, or ſlate colour.

It inhabits Europe, not farther north than Sweden; lays four or five pale blue eggs.

In England the Hedge Sparrow is very common all the year, in France it is a bird of paſſage; it comes in autumn, and leaves that country in the ſpring. They travel in companies, and alight in hedges, flying from buſh to buſh. The neſt is built very low in garden hedges, or ſome ſmall buſh, or ſometimes upon the ground; it is compoſed of moſs on the outſide, of wool, and horſehair within; it lays five eggs of a light blue colour, and without ſpots. If a cat or any other animal ſhould approach its young, like the Partridge,

tridge, or Lapwing, in order to miflead, it would
flutter juft before its enemy, counterfeiting lame-
nefs, till it had drawn it to a confiderable diftance
from the neft.

During the breeding feafon it has a remarkable
flirt with its wings.

The Hedge Sparrow in the feverity of winter,
approaches barns, and threfhing floors, in order
to pick up corn, but this is not its natural
food, it feeds on chrifalis's, plant lice, and other
infects.

The Hedge Sparrow is not fufpicious, and may
be caught in almoft any trap, he may eafily be
tamed, and by fome is valued for his fong, though
it is not any thing extraordinary, it is plaintive,
and he fings often at a feafon of the year when
other birds are filent. It is generally towards the
evening that they fing the moft; they begin with
the firft froft, and continue till a little time in
the fpring.

A gentleman in France put a Hedge Sparrow
into an aviary, with Canary Birds, Linnets, and
Goldfinches; a Canary Bird feemed to take an
affection to it, and never quitted it; they were
taken out, and put into a cage together, but made
no neft.

THE

THE REED WARBLER'S

Plumage is a reddish brown, beneath of a light reddish colour, the head spotted.

It inhabits Europe, and the bog rushes in Sweden.

It is the size of a Wren, very lively, quick, and gay; it has a song, which though but little varied, still is not disagreeable, and may be improved. It sings in the winter; in the spring it returns to the woods, and builds a nest there of green mofs, lining it with wool, and lays five eggs. The young ones are easily reared, and their little song, and sprightliness, is very interesting. A gentleman who had brought up some in an aviary, gave all his birds their liberty in the spring; these were the last that left it; they were afterwards frequently attacked by wild birds of their species: when thus persecuted, they flew to the window cills, as it were for protection, and there defended themselves, ruffling their feathers, and fighting the wild ones like game cocks.

THE KRUKA WARBLER

Is dusky above, whitish beneath; tail-feathers dusky; each outer one striped with white on the outer web.

It inhabits Sweden, and all parts of Ruffia, but not Siberia; and sings in the night.

This

This bird is heard almoſt inceſſantly in the ſpring; it is often ſeen to riſe ſtrait above the hedge, whirl round in the air, and deſcend, ſinging always the ſame lively little ſong, which it continually repeats, ſo that it has been called the prater.

One ſees it conſtantly flirting, reſtleſs, entering into the buſhes, hopping about them, without ever perceiving it ſtill for one moment. It makes its neſt in hedges by the road ſides, in places the moſt concealed, near the ground, and even in tufts of graſs which grow in the bottoms of buſhes.

It feeds upon inſects, eſpecially thoſe caterpillars that are found upon the leaves of ſhrubs, and buſhes: as it frequents our gardens, groves, and the neighbourhood of houſes, it is already in part familiarized, and may be eaſily tamed; is ſometimes kept in a cage for the ſprightlineſs of its ſong; it is neceſſary to let it have water to bathe itſelf, otherwiſe it would die in moulting.

PETTY CHAPS.

The plumage greeniſh aſh colour, beneath yellowiſh; ſtomach inclining to white, or ſilvery; tail duſky; inſide of the mouth red, eyebrows white.

Inhabits Europe as far north as Sweden.

This

This sprightly little bird appears early in the spring; as soon as the leaves begin to open, and the bloſſoms to appear, they diſperſe themſelves through the country: ſome frequent our woods and groves, and others our gardens; they animate every country ſcene, by their ſprightlineſs and gaiety. There is nothing pleaſing in their plumage, which ſeems to be the caſe with moſt of thoſe birds whoſe manners are ſprightly, or whoſe ſong is melodious.

They frequent gardens, and fields that are ſown with peas; they build their neſts upon the riſers that ſupport theſe plants, and are continually going into and out of them; they are in conſtant action, ſporting with, watching, and purſuing one another; their little quarrels ſeem without reſentment, and are always ended by a ſong.

Notwithſtanding their gaiety, they are far from inconſtant; and the affection of the parents towards each other, ſeems to continue after the breeding ſeaſon is over.

Whilſt the female ſits, nothing can be more attentive and aſſiduous than the male. The neſt is compoſed of dry graſs, filaments of hemp, and horſe-hair. The female lays five eggs; ſhe forſakes them if they are touched, and it is impoſſible to deceive her by putting others in their

room,

room, fhe immediately difeovers, and throws
them out.   Yet it is faid, that in the neft of
this bird the cuckow frequently lays its egg,
and that fhe attentively fofters its young.

The Petty Chaps is of a timid difpofition, flies
away from birds as fmall as itfelf, and is very
juflly afraid of the Butcher Birds, its for-
midable enemy.   The moment the danger is
paft, it feems entirely to forget it, and all its
former gaiety returns.

It generally fings perched among the moft
tufted foliage of a tree; fometimes for a moment
it fhews itfelf on the outfide of a bufh, and in-
flantly hides itfelf again amongft the leaves.

In the morning it bathes itfelf in the dew upon
the leaves, or after flight fhowers in the fummer.

As they feed principally on infects, they leave
us in autumn, though there are fome berries upon
which they will feed.

SEDGE BIRD.

Plumage, afh colour, white beneath, with white eye-brows.
Head brown with dufky ftreaks, tail brown, and circular
when fpread; toes fulvous and yellowifh.

Inhabits Europe, is frequent in Ruffia, and Siberia, among the willow thickets, near the
rivers,

This

This bird is sometimes called the willow Nightingale, because it sings in the warm and clear
nights in the spring; it makes its nest among the
reeds, in bushes, amongst marshes, or in the thickets
that over-hang the water. The nest is composed
of straw, dry grafs, and lined with horse-hair,
and constructed with great ingenuity: The female
lays five eggs, of a dingy white, spotted with
brown.

The little ones, before they are feathered, will
throw themselves out of the nest if any body
touches, or even comes very near them, though
the nest be built immediately over the water.

This bird darts from among the reeds to pur-
sue the dragon fly, on which it feeds, and it drives
away all other little birds from its immediate
neighbourhood.

It is an entertaining mock bird; it fits concealed
in willows, or reeds, and imitates in a pleasing,
but hurrying manner, the Swallow, Sky-Lark,
and House-Sparrow. It seems to leave us before
winter.

The WHITE THROAT.

Plumage ash colour above, white beneath, the outer tail
feather half white all the length, the second white at the
tip.

A nest

A neft of the White Throats was found on a plum-tree about three feet from the ground, in the form of a cap, compofed of mofs, interwoven with bents of dry grafs. Sometimes it is built entirely of grafs coarfer without, and finer within; White Throats lay five eggs of a greenifh grey, with red and brown fpots.

In countries where figs, and olives grow, it feeds on thofe fruits. It frequents our gardens in fummer, and leaves us in winter. It is a fhy, and wild bird, and fings with an erected creft.

THE EPICUREAN WARBLER.

The plumage rather dufky, white beneath; the breaft fpotted with afh colour. Inhabits Europe, but moftly the fouthern parts.

It migrates into the more northern parts later than other birds of this Genus, and leaves them fooner in the fummer; they are found in Germany, and Poland, and as far north as Sweden; in autumn they return to Greece, and Italy; and in winter probably go to ftill more fouthern climates. Their manners, and food, feem to change with the climate: in the north they feed much on infects, in warmer countries chiefly on fruit during the feafon, as grapes and figs.

In

In fouthern countries they arrive in flocks,
in temperate climates they are difperfed in the
woods; their nefts are fo artfully concealed, that
it is very difficult to find them; they perch high
in trees, and warble very prettily.   The plumage
is not beautiful, but they are a moft delicate food,
and almoft equal to the Ortolan.   They fly
by jerks; they walk, but do not hop, and run
along the ground in France among the vines.
They leave France about Auguft, migrating in
little companies of five or fix.   They are caught
with noofes, by nets, and by means of a looking-
glafs.

They are taken in great numbers in Provence,
and the iflands in the Mediterranean.   When
Cyprus belonged to the Venetians, thefe birds
were an article of commerce; 1200 jars filled
with them, preferved with vinegar, and aromatic
herbs, have been annually fent to Venice, and
nothing can be more delicate when they are
fattened upon grapes and figs.

The   W H E A T   E A R.

The back grey, tinged red; the forehead white; a black
band from the bill to the hind part of the head; the extreme of the body, and upper part of the tail white, the
tail-feathers black at the tips.

The female has not the black band near the
eye.

It

It Inhabits warm, and ftony places in Europe; making its appearance when the cold winter nights are paft.

This fpecies is found from the fultry climate of Bengal, to the dreary regions of Greenland; it is migratory in the temperate and frigid zones. In Greenland it frequents rivulets, and feeds on worms among the graves; for this reafon it is abhorred by the natives. In Sweden the farmers confider it as the harbinger of fpring, and that it points out to them the time they may with fafety fow their corn.

Wheat Ears are very common in England; they come in the fpring from March to May; the females arrive a fortnight before the males: They frequent commons, and warm downs, and the fides of hills, thofe efpecially that are fenced with ftone walls, perching upon the little tufts of earth.

In ploughed grounds they follow the furrows to pick up worms, on which they feed; when difturbed they do not rife high, but fkim with a fhort, but rapid flight, near the furface of the ground; and foon alight. In flying, the Wheat Ear difcovers the white part of his tail, and the white feathers at the end of his body. He is often feen in barren, and in fallow ground, flying from

ftone

9

ftone to ftone, feeming to avoid hedges and bufhes, upon which he alights much lefs feldom than on ftones.

The beak is fine at the point, but wider at the bafe, confequently well conftructed for feizing, and devouring infects, on which they dart continually. They almoft always keep on the ground, and if difturbed, perch only on low bufhes. When they alight, they twitter, and flirt their tails. In ground lately ploughed, or near little tufts, or under ftones in fallow land, or between the ftones of which they make fences in fome countries, they build their neft. It is curioufly conftructed of mofs, or fine grafs on the outfide; of feathers and wool, rabbits down or fur, and horfe-hair in the infide. It is remarkable for a kind of fhelter placed above the neft, and faftened to the ftone, or hillock, under which the neft is made. The female lays five or fix eggs; and fits fo clofe as fometimes to lofe the feathers from her breaft. The male attends her with great affection, bringing her flies, and ants, and always keeps near the neft. If he obferves any perfons approaching, he flies before them, alighting every now and then, as though to divert them from the neft; and when he judges them at a fufficient diftance, he takes a compafs, and returns

turns to his fituation. Wheat Ears feem impatient of cold ; and if any fevere frofts happen after their arrival, it is fatal to many. They prefer high, and dry fituations. When they are fat, they are delicate food. They are taken in great numbers in hair noofes, by the fhepherds about Eaftbourn in Suffex. The fhepherds cut out a turf, and lay it along by the fide, and over the trench, made by the removal of the turf ; leaving only a little hollow, in which the noofe is placed ; the Wheat Ear with a view to find worms and perhaps to hide itfelf, goes into the noofe. The appearance of a bird of prey, or the fhadow of a cloud, are fufficient to frighten him into this fuppofed fhelter ; the number taken in that neighbourhood every year, amounts to 1840 dozen : the reafon why they are fo numerous there is becaufe that fituation abounds with a certain fly, which for the fake of the wild thyme frequents the adjacent hills. They migrate in Auguft or September, and go in little flocks ; they are naturally folitary, for they difperfe as foon as they arrive, the male and female only affociating together. They feed on infects, and earth worms.

The WHIN CHAT.

Plumage a rufty brown fpotted with black, a white line over each eye, a broad patch of black beneath, and covering the temple in the male, of brown in the female; two white fpots on the wing. The throat and breaft yellowifh, two middle feathers of the tail black, the reft white at the bottom, black at the tip.

It inhabits Europe as far north as Spitfbergen.

The Whin Chat feldom perches, but is moftly upon the ground, upon little hillocks, in fallow lands. The female lays four or five eggs, and builds her neft at the bottom of a bufh, amongft the roots, or on the ground, where it is fheltered by a ftone. This bird is of a wild nature, it arrives and departs with the Stone Chatter, and frequents mountainous fituations; it feeds on flies and other infects, and when fat is as delicious food as the Ortolan.

The STONE CHAT.

Plumage grey, red beneath, a white band under the throat, the fpaces between the beak, and eyes, black.

This fprightly, active bird is fcarcely ever quiet; almoft continually hopping and fluttering from bufh to bufh, he only repofes for a few moments, and even then he feems fpreading his wings, as though meditating another flight. He rifes into the air by fhort and fudden efforts, and falls, turning round like a wheel.

PART VI.   G   T ·

The Stone Chat never flies very high, or perches upon trees, but generally alights upon the highest fpray of a hedge, or a bufh. He is eafily taken with bird-lime, from his habit of preferring any fingle twig which projects beyond the reft. He frequents dry fituations, high fields, or extenfive commons : fometimes he perches on the ears of Indian wheat in the fields, or on the higheft props in the vineyards.

This bird builds its neft on waftes, or commons, at the bottom of a bufh, amongft the roots, or under the cover of a ftone, or fometimes fixes it to the fide of a rock, but fo fufpicious is the character of the Stone Chat, that it retires there in a moft artful and cautious manner, as though afraid of being feen : it never goes immediately into the neft, but firft flies into a bufh at fome diftance ; when it leaves its neft it walks a little way from it, and comes out of a neighbouring bufh, fo that if you fee the bird enter haftily any little thicket, it is not there, but at fome little diftance, you muft feek for its neft.

The female lays five or fix eggs, both parents feed their young with infects, which they are bringing them almoft perpetually. Their cares, and attentions, feem increafed when their young are able to leave the neft, they are repeatedly calling to, and still continue to feed them for several days.

In confinement, the character of this bird seems quite altered; he becomes dull and stupid, incapable of learning any thing; he is kept alive with great difficulty, and to no purpose.

A Stone Chat that had been taken shewed great reflexion; the cage in which it was confined was placed in a garden amongst shrubs, in broad day, and the door opened; it instantly flew upon the outside of its little prison, and there remained a minute before it attempted a second flight, distrusting as it were the appearance of liberty.

In their natural state they suffer you to approach near, flying only to a little distance, not seeming to suspect the fowler's intention.

There is a bird in Provence which lives very much upon ants, for which the Stone Chat has been mistaken; but the Ant-eater seems to be a solitary bird, frequenting only decayed houses and ruins. In cold weather, he places himself on the tops of chimnies for warmth; he frequently lies at the opening of an ant hill, stopping it with his body, so that the ants who attempt to get out are engaged in his plumage; he then flies away to a bare spot, and shaking his feathers, the ants fall out upon the ground, and become his prey. It is most pro-

G 2

bable that this bird is not a Stone Chat, but he may belong to another fpecies of the Motacilla.

The Stone Chat, when fat, is delicious food, and nearly equal to the Ortolan.

The BLACK CAP.

The crown of the head is black, hind part of the head pale afh colour; back and coverts of the wings greyifh olive; breaft and ftomach light afh colour: crown of the female dull ruft colour.

It is one of the fmalleft of this genus.

This bird has not the black crown until after the firft moulting, before that time the young birds refemble young Fig Eaters; from whence it was once ftrongly imagined, that Fig Eaters were fometimes metamorphofed, or changed into Black Caps.

Its fong is very pleafing, not unlike that of the Nightingale, and it is of much longer continuance; for, many weeks after that delightful warbler is filent, the fong of the Black Cap ftill make the groves mufical.

Their notes are eafy, natural, and pleafing; not confifting of any great variety, yet they are agreeable, and feem expreffive of the happinefs and tranquility of the furrounding fcene. The male has a thoufand attentions for the female whilft whilst she sits; indeed, he shares that care with her, and is continually bringing her flies, worms, and ants.

The nest is built near the ground, in a bush carefully concealed, and contains four or five eggs, of a greenish colour, with light brown pots. It has been found in a spruce fir, composed of goose grass, moss, and wool, lined with horse hair.

The young ones grow very fast, and when they are slightly fledged, they leap out of the nest, and abandon it, if any persons approach.

This bird arrives in the spring, and if after its arrival occasional frosts should make the insects disappear, it has still a resource in several kinds of berries.

The Black Cap may be tamed, and in confinement few birds can be more interesting; it shews the most engaging affection for its master; at his approach it flies against the wires of the cage to meet him, and by its gentle notes, and the fluttering of its wings, expresses gratitude and attachment.

The young birds who are brought up in confinement, if they have opportunities of hearing the Nightingale, acquire its song; but in September, which is the time of their migration, they

G 3

shew

shew all thofe fymptoms of uneafinefs, which migratory birds generally feel at that feafon.

The Black Cap is found commonly in Italy and France, fometimes in England, and as far north as Sweden.

The RED START.

The head and back are of a bluifh grey; the checks and throat, black; the ftomach and tail red, except the two middle tail feathers, which are brown.

The fong of the Red Start has neither the compafs or the variety of the Nightingale's, but it has an expreffion of tendernefs and melancholy, which is exquifitely pleafing; for this reafon, and this only, it is called in France, the Nightingale of the Wall, for it has neither the form, nor the habits of the Nightingale, nor its plumage.

The Red Start arrives in the fpring, perching upon towers, upon the roofs of houfes, and on chimnies, from whence is heard his foothing fong. He always feeks the moft inacceffible fituations; he is found too in the fhades of the gloomieft forefts; he flies lightly, and when perched, has a little cry, which he repeats, moving his tail with a kind of tremulous motion, not up and down, but from one fide to the other, like a dog when it fawns. This bird prefers mountainous countries.

They

They build their nests in holes of walls in towns, and in the hollows of trees in the country, or the fissures of rocks. The female lays four or five, or six blue eggs ; the young are hatched in May. The male during the time of incubation sings from the point of a rock, or the top of some building near the nest ; it is, especially, at the dawn of day, and early in the morning that he repeats his song.

The Red Start is a shy, suspicious, and sullen bird, apt to abandon its nest if it be observed, and it is said to forsake the eggs, and even the young if they are touched ; if caught, it cannot be tamed, but obstinately refuses food, and dies.

These birds from the nest may be reared in confinement ; they will sing delightfully in the day, and sometimes in the night, and improve their natural song if placed among better songsters than themselves. The tame Red Start is fed with crumbs of bread, and the same composition that is given to the Nightingale.

When unconfined they feed on flies, spiders, chrysalids, ants, berries, and tender fruits ; in Italy on figs.

In their migrations they preserve their solitary character, they go alone, and are never seen in flocks.

The GREY RED START.

The beak and quill feathers aſh colour, the ſtomach and tail feathers ruſt colour, the two outer feathers aſh colour.

The Grey Red Start is a migratory bird, it arrives in May, and immediately retires into the woods to paſs the ſummer: it builds in little buſhes near the ground; the neſts are made with moſs on the outſide, and lined within with wool and feathers, reſembling a globe, with an opening towards the eaſt. According to Buffon the female lays five or ſix eggs; according to Pennant nine.

In the morning it comes out of the woods, ſojourns there during the heat of the day, and in the evening appears again in the neighbouring fields, ſeeking for worms and flies; it retires to the woods at night to rooſt. The Grey Red Start has no ſong, only a little twittering note; in general he is very ſilent; he often perches upon a ſingle twig, which grows out beyond the root, flirting his tail like the Red Start. When fat, he is delicious food: his flight is very ſhort, only from buſh to buſh. Theſe birds go away in October; they are ſeen ſome days before to follow one another along the hedges.

Some

Some grey Red Starts have a black band round their necks, and their plumage is more brilliant; thefe are fuppofed to be the males.

The BLUE THROATED WARBLER.

The breaft ferrugineous, with a blue band, the tail fea-thers dufky, ferrugineous towards the bafe.

This bird, except in colour, very much refem-bles the Robin Red-breaft; it frequents the bor-ders of woods, marfhes, and moift meadows, willow beds, and reeds, and with the love of folitude that diftinguifhes the Red-breaft duri. part of the year, it feems to have the fame famili arity with man ; for, after fpending the fummer in thefe retired fituations, in the autumn, it fre-quents hedges, avenues, and gardens, and fuffers you to approach near enough to beat it down with a farbacan ; a farbacan is a hollow tube, or cane, through which peas, &c. are blown.

Seldom more than a pair of the Blue-throated Warblers are feen together. At the clofe of the fummer they refort to corn fields, and fome-times make their nefts there, but more generally upon willows, ofiers, and plants, that grow in

G 5   marfhy

marfhy places; their nefts are compofed of grafs, and fixed to the forks of branches.

In the breeding feafon, the male rifes ftraight in the air a little way, finging; he turns round, and fo alights again upon the branches with a great deal of fprightlinefs and gaiety. The Blue Throated Warbler fings in the night, and fome think the fong pleafing, by others it is little efteemed; but it has been obferved in feveral fpecies of birds, that the fong of thofe of the fame fpecies difiers at different feafons of the year.

It loves to bathe itfelf, and keeps very much near water; it feeds on worms and infects, and runs very nimbly, flirting its tail.

The fine colours of this bird lofe their luftre when he is in a ftate of confinement.

This fpecies is not very common; it is found i the mountainous parts of Europe, in the Alps, the Pyrenean Mountains, and in Sweden.

The BLUE RED-BREAST.

Blue above, red beneath, the ftomach white, the qui feathers black at the point.

Its wings are very long, and its flight rapid; it is of a gentle difpofition, feeds on infects, and makes its neft in the hole of a tree.

The

The Blue Red-breaft inhabits the Ifland of Bermuda and South America; it is harmlefs, and familiar.

It is in America what the Robin Red-breaft is in England. The Blue Red-breafts frequent fields where maize and mulleins grow, perching on their ftalks in order to pick off the flies; frequently they are feen on rails, fpringing at flies as they pafs.

The CERULEAN WARBLER.

Blue above, white beneath; the wings and tail are black, the outer tail feathers are white.

It inhabits Pennfylvania.

The females begin in April to build their nefts; the outfide is compofed of lichen, the infide is lined with th. fine down of plants, and between the mofs and the down is a layer of horfe-hair; the form of the neft is cylindrical, clofe at bottom, open at the top.

The SYBIL.

Plumage blackifh, white beneath, breaft a dufky red, a white fpot on the wings.

It inhabits Madagafcar, and fings d.. .. tf.ily.

G 6   The

THE RED-BREAST.

Plumage a kind of olive colour, the throat and breast a deep orange red.

It inhabits Europe, feeds on the feeds of the spindle tree, sings excellently, is quarrelsome, and consequently solitary; its fearlessness of mankind has occasioned it every where to be distinguished by a familiar name. The English call it Robin Red-breast; the Danes, Tommi Liden; the Norwegians, Peter Renfmad; the Germans, Thomas Gierdet. In several parts of Europe Red-breasts are birds of passage; in general they spend the summer in the woods, and at the close of the year frequent the habitations of men. They build their nests near the ground, upon the roots of young trees, or amongst plants strong enough to support them, in the thickest coverts, or most concealed holes of walls. The nest is made of moss, leaves, and horse-hair, and lined with feathers; they often cover it with leaves, leaving only under the heap a narrow and oblique entrance, which the female conceals with a leaf, when she goes out. She lays from five to seven eggs, of a brownish colour, or dull white, sprinkled with reddish spots. Whilst she sits the male warbles in the

woods

woods his tender and harmonious song; he seems entirely engroffed with his mate, and suffers no birds of his own species to live very near. There is a common proverb, that two Red-breasts are never found in one bush.

Red-breasts seek the shade, and prefer moist situations: in the spring they feed on worms and insects, which they take with great addrefs; they flutter like a butterfly about a leaf, on which they perceive a fly; on the ground they spring forward with little hops, and dart on their prey, flapping their wings. In the autumn they eat berries and fruits.

There is no bird that sings more early: he is awake the first in the woods, and begins his song by the dawn of day, and he is the last that we hear, or see flutter about in the evening; besides, his song continues the greatest part of the winter and the spring.

No bird is so easily taken by call-birds or by traps; it is scarcely possible to make any noise which shall not awaken the curiosity, and engage the attention of all the Red-breasts that are near; and if any bird-lime twigs be placed, they fly upon them; should any one effect his escape from that snare, he makes a little noise, which occa-

sions

fions the reft that are not taken immediately to
fly away.

The young have not the red feathers on their
breaft until after their firft moulting. About the
beginning of October, many of them prepare
to leave France; they do not go in flocks, but
preferve in their migrations their folitary character; they return to France in April. A gentleman in that month had taken feveral in nets for
three days, one after another; on the fourth day,
the morning being fine, he expected his ufual
fuccefs, but was difappointed, not one bird could
be found.

However, many Red-breafts remain all the
year in France; in winter they become very
tame, and when the ground is covered with fnow,
approach houfes, and even fly into the rooms if the
windows be left open; they difcover an affecti on-
ate familiarity, coming to pick up crumbs on the
table. In a monaftery, a Red-breaft was kept
in one of the cells; in two or three days he
feemed quite at his eafe: he fpent his time
there very happily all the winter: in the fpring he
ftruck with his beak againft the window, as though
he wifhed to go, and the window being opened he
withdrew. Sometimes they have been known to
come for years together to the fame houfe, and

to

to fit and fing on the fhoulders or chairs of thofe who fed them. A gentleman, in Staffordfhire, found a Red-breaft one morning in his chamber; it fat and fang on a chair ; he opened the window to let it go out, but the bird feemed to have conceived an affection for him, and flew after him down ftairs, fed on fome crumbs whilft the gentleman was at breakfaft, amufing him with his fong, and flying about the houfe all day. At night, the little fociable bird followed him up ftairs when he retired to his chamber, and the next morning he was awakened by the Robin fluttering over his face ; the bird then retired to his chair and began to fing, and in this manner continued to live for fome time with the gentleman, though every day he might have gone out, the windows being left open with that view. In a tame ftate, Red-breafts will eat almoft any thing, bread, or fmall pieces of meat, or grain.

As the Red-breaft lives moftly on infects in fummer, it is a bird difficult to bring up in a cage, though naturally tame : his beak is flight, like that of birds that feed chiefly on infects.

In autumn they are very fat, and delicate food.

In England the Red-breaft is held in peculiar efteem. The old, and favourite ballad of the Children in the Wood, may not have contributed

a little

a little to the tendernefs with which it is treated
by the playful fchool-boy; and the laft kind
offices which they are reprefented in that fimple
fong to have performed to the little innocents
who died in the wood, in covering their bodies
with leaves, naturally enough tends to excite in
young and feeling minds, acts of forbearance,
and fentiments of affection.

THE WREN.

The plumage is grey, the wings waved with black, and
afh colour.

The bill is very flender, pointed, and fcarcely bowed;
the wing feathers on their outer webs are croffed with many
little dufky bars; the firft, fecond, and third are fpotted
with white between the bars.

The tail is rounded and croffed with dufky, blackifh lines.

The Wren is a fprightly little bird, frequenting villages, and the neighbourhood of towns ..
the approach of winter, and chanting, efpecially
towards the evening, with a clear voice, his
pleafing, animated fong.

He fometimes fhews himfelf for a moment upon
a heap of dry wood, the next inftant enters it,
and difappears. For a moment he is feen upon the
edge of the thatch, but quickly conceals himfelf
                                        under

under it, or in fome hole in the wall. As foon as he comes out, he frifks amongft the thick branches of the neighbouring bufhes, always raifing his little tail.

His flight is quick, and irregular, and he flaps his little wings fo rapidly that you do not fee their motion.

He preferves all his gaiety during the cold and gloomy feafon of winter, and fings his little lively fong, in which, notes, like the words *fideriti*, *fideriti*, are often repeated, with more than ufual animation and perfeverance, when the falling fnows and chilling cold feem to make all nature filent and torpid.

that feafon he frequents our court-yards, particularly the wood houfes, feeking amongft the fticks and under the banks, under the roofs of our outhoufes, in holes of walls, and even in the mouths of wells, the chryfalids, and the dead bodies of infects. He frequents the fides of ftreams that do not freeze, retiring occafionally into the hollow parts of decayed trees, where twenty have been found crouded together.

Although they are neither fhy, nor timid, they are not eafily taken; they are fo nimble, and fo mall, that they elude the attempts of their en- nemies.

In

In the spring the Wren frequents woods, where it builds its nest under thick and leafy branches, at no great distance from the ground, sometimes in a mossy bank, or in the projecting margin of a little stream, under the trunk of a tree, against a rock, and sometimes in the roofs of the solitary cots of faggot-makers in the woods. The outside of this curious structure is composed entirely of moss, and neatly lined with feathers. It is almost round, and so uncouth on the outside as to resemble a moss-grown clod, and excite of course no attention. A little entrance is left on the side; the Wren lays from ten to eighteen small eggs, of a dingy white, spotted at the larger end with red.

If she perceive that her nest is discovered, she immediately forsakes it.

Sometimes the field-mouse takes possession of a Wren's nest, but whether it be first deserted, or whether the mouse destroys the young, is not known.

Like the Robin the Wren sings late in the evening, and early in the morning.

It inhabits every part of Europe, and continues with us in England all the year.

THE

THE GOLDEN-CRESTED WREN.

On the crown the feathers are orange coloured, bounded on each side by black; plumage above a yellowish green, a reddish white beneath; the wing coverts dusky, crossed with two white bands; quills and tail dusky, edged with pale green.

This bird is the smallest that is known in Europe; it escapes through the meshes of our common bird-nets, and the wires of our closest cages. If we put it into a room, in time it disappears through some little unperceived opening. In our gardens it soon eludes our sight, and no wonder, since there are few leaves under which it may not sit concealed. If you attempt to shoot it, with a view to preserve the figure of the bird, the finest sand is sufficiently coarse for the purpose, so minute is the beautiful little object of your pursuit. Its cry resembles that of the grasshopper, and it scarcely exceeds the grasshopper in size.

The female lays six or seven eggs in a curious nest, like a hollow globe, the outside is of moss, the inside is lined with the finest down, and the entrance is a little hole in the side. She builds generally in woods, sometimes in our gardens, in pines or yew trees.

The

The fmalleft infects are their food. In the fummer, they take them on the wing; in the winter, they feed upon their larvæ, their chryfa-lids, and upon little worms; occafionally they eat fmall feeds, particularly the feeds of fennel, and they turn over the earth or decayed wood which is in the hollow parts of willows, proba-bly for the larvæ of infects.

They frequent oak trees, feemingly in prefe-rence to any others. They are very fprightly, and in perpetual motion, running along the branches in every direction, fometimes with their backs downwards like the Titmoufe, fearching for infects in every little cavity of the bark. The Crefted Wren is found in every quarter of the globe, and ftays with us in England all the year. Its fong is faid to be delightful, but weaker than that of the common Wren ; it has been feen fuf-pended on the wing for a confiderable time over a bufh in bloffom, finging melodioufly.

The YELLOW WREN.

The plumage is an olive green, the eye-brows yellowifh, the wings and tail are brown, edged with yellowifh green.

Its food is flies and other little infects During the fummer it inhabits woods, making its neft in the moft concealed parts of bufhes, or in thick grafs ;

grafs ; the neft in its conftruction, and the materials employed, refembles that of the common
Wren, and probably this form, as it feems peculiar to our fmalleft birds, has been fuggefted by
the wife Author of Nature, and of Inftinct, becaufe in thefe cold climates, their warmth might
be infufficient for the purpofe of incubation.
This little bird is much attached to its neft, and
in this refpect widely differs from the common
Wren. A gentleman having found a neft belonging to this little bird, by taking away the
eggs as they were laid, occafioned her to produce thirty in fucceffion ; he then took pity
upon her, and left a fufficient number for her to
fit upon. In autumn it quits the woods, and
frequents orchards and gardens. Its fong continues all the fpring and fummer, it is full, fweet,
pleafing, and continued.

In France, they arrive in the month of April,
in little companies of 14 or 15, but they foon feparate, and pair. If any very fevere cold fhould
happen after their arrival, it is fatal to them,
and they are found dead upon the ground.

GENUS 77. PIPRA.

The beak is fhorter than the head, towards the bafe
flightly three-fided, ftrong, hard, a little incurvated, or
bowed.

In

In moſt ſpecies the middle toe is connected to the outer toe, as far as the firſt joint.

Manakins, in general, are very little, and very pretty birds. They are not much known, as they principally inhabit the extenſive woods in the warmer climates of America, and ſeldom quit them to reſort to expoſed ſituations, or to approach the habitations of men. Their manner of flying is low, and rapid, and ſeldom to any conſiderable diſtance.

They perch not at the tops, but on the middle, or lower branches of trees, feeding upon wild berries, and ſometimes on inſects. In general they are ſeen in little companies, of eight, or ten, moſtly of the ſame ſpecies, but ſometimes with other birds, even of a different genus: it is in a morning that they are thus collected together, they ſeem to enjoy then the ſweets of ſociety, and expreſs their chearfulneſs by a gentle and pleaſing warbling. In the heat of the day they ſeparate, retiring from the painful ardour of the noon-day ſun, into ſhady retreats. They prefer riſh and verdant ſituations, though they do not haunt marſhes, or the banks of ſtreams.

The ROCK MANAKIN

Has an erect creft, with a purple margin. The plumage is of a faffron colour, and the coverts of the tail are fquare, as though cut off.

This bird is the fize of a fmall pigeon, and very beautiful; the plumage, though of one colour, being very regularly difpofed. When young, its feathers are brown, afterwards of a ruft colour; it is only by degrees that it acquires that beautiful orange, which diftinguifhes this bird in its more advanced age.

It is an inhabitant of South America, and is found in various parts of Surinam, Cayenne, and Guiana, in rocky fituations; but no where fo frequent as in the Mountain Luca, near the River Oyapoc, and the Mountain Courouaye, near the River Aprouack: they build there in deep chafms of the rock, and in large caverns and recefles, where the fun never penetrates. Their nefts are of a very coarfe conftruction, made like that of the wild pigeon, principally of dry fticks.

It is probable fince they are fo fond of an obfcure retreat, that their eyes are capable of confiderable contraction, and dilatation, like thofe of the owl, because they are feen in the day, par-

ticularly

ticularly the male birds, in the neighbourhood of thefe caverns.

They are very fhy, and difficult to be fhot, for as foon as they perceive the fowler, they fly with great rapidity, though to no very great diftance; their flight is low.

The female lays two eggs : as fhe advances in years her plumage brightens, fhe difcontinues breeding, and it is afferted, in time, acquires fo entire a refemblance to the male, as not to be eafily diftinguifhable; this too has been faid to be the cafe with feveral of the Gallinaceous birds.

Though naturally wild, they have been tamed fo far as to go about with the poultry, to which tribe of birds they bear a confiderable refemblance.

THE BLUE BACK MANAKIN,

Has a beautiful crimfon creft, the plumage on the back is blue, on other parts black.

It inhabits Brazil, and Cayenne.

When young it is green, except its crimfon creft. It frequents woods, feldom wandering far, and fitting moftly perched upon the lower branches of trees, where it finds thofe feeds, and infects, on which it feeds. Blue Back Manakins are frequently in fmall flocks, and utter in

concert

concert a little note, at fhort intervals, though not pleafing in itfelf. It is often grateful to travellers, bewildered in the vaft forefts of Guiana, for when they hear the cry of thefe birds, they are certain that water is near.

The BLACK CAP MANAKIN.

The crown and hind part of the head is black; the chin, fore-part of the neck, and under parts of the body white, thefe white feathers paffing round the neck form a ring. The reft of the plumage is black, except a white fpot on each wing.

Like other Manakins they occafionally live in little companies, but do not mix with birds of different fpecies.

The cry they utter refembles that which is made by the little inftrument we ufe in cracking nuts, from which circumftance Mr. Buffon has given it the name of the Nut-cracker. They are very fprightly, and active, feldom in a ftate of repofe, but frifking from branch to branch, without flying to any confiderable diftance.

A variety of this genus, without the white fpot on each wing, frequents the neighbourhood of ants nefts; thefe little birds are obferved to fpring occafionally from the ground, uttering a fingular cry, probably in confequence of their legs being ftung by the ants.

ii

THE LITTLE MANAKIN.

The plumage is grey, the head black, fpotted with white.

It inhabits India, and is about the fize of a Wren.

THE WHITE CAP MANAKIN.

The head is white, the general plumage a grey brown.

It inhabits Brazil and Surinam, frequents reeds, and fings pleafingly.

THE TUNEFUL MANAKIN.

The forehead is yellow, the crown of the head and neck blue; the reft of the plumage above, dufky, and black; the lower part of the back, breaft, and thighs, orange.

It inhabits St. Domingo, where it has acquired the name of Organift, from its fong, which forms a compleat octave, in the moft agreeable fucceffion of notes.

It is probably the fame bird mentioned by Du Pratz; its notes, he fays, are fo varied, fo melodious, and warbled in a ftile fo tender, and expreffive, that thofe who have heard it are lefs captivated with the fong of the Nightingale.

Mr. Du Pratz obferves, that it fang perched on an oak near the houfe where he lived.

It

It has been remarked that this interesting songster sings for near two hours, with scarcely any intermission; and after a respite of about two hours, begins again its delightful melody.

GENUS 78.   PARUS.

The bill is strait, strong, hard, and sharp pointed. The nostrils are round, and covered with bristles, which are reflected or turned back over them, from the base of the beak. The tongue appears as though cut off at the end, and is terminated by three or four bristles.

The Titmouse Genus, seems to have a near affinity with several birds of the order Picæ, particularly the Woodpecker, Creeper, &c. Like them they often build in hollow trees, and creep along the under side of the branches, or run up and down the trunk, seeking for insects, in the irregularities of the bark. Some species of the Titmouse, like the Oriole, make a pendulous nest. Indeed, from several circumstances in their general habits, they seem to constitute a miniature resemblance with several of the Picæ.

Most birds of this genus are apparently weak, because they are generally diminutive in size; but they are at the same time sprightly, active, and spirited. In constant action, flitting from tree to tree, hopping from one branch to another,

H 2   climbing

climbing the bark, or running up walls, they
fix, and fufpend themfelves in every poffible
form; fometimes with the head downwards, feeking for worms, infects, their larvæ, chryfalids, or
eggs. They feed too on feveral kinds of feeds,
but inftead of breaking them like Linnets, and
Goldfinches, between the mandibles of their beak,
they generally place them between their feet, and
pierce them with the point of the bill. It is curious to fee them feed in this manner upon hemp-
feed; they will pierce the fhells of nuts and almonds, and if a nut be fufpended at the end of a
ftring, they will fix upon it, ftrike it with the bill,
and fuffer themfelves to vibrate with the ftring,
without once loofing their hold, or ceafing to
peck the nut.

The mufcles of their neck are very ftrong,
and their fkulls are thick; this explains in part
their manœuvres, but to account for the whole,
we muft fuppofe that the mufcles of their legs and
feet are very ftrong.

Towards the clofe of autumn, they frequent
the neighbourhood of our habitations, feeding upon
feeds, and the infects which the cold of winter
has deftroyed. They fearch too for the bodies of
dead birds; and if they find any alive, entangled
in traps, or weakened by difeafe, they take an ad-
vantage

vantage of their diftreffed and defenceles fituation, and though they fhould be of their own fpecies, they pierce their fkulls, and feed upon their brains. This fpecies of cruelty feems implanted in their nature, and to influence them when they are not urged by want; for in a cage, when they are fupplied with food in abundance, they will ftill practife this unneceffary act of violence.

Their food is very various. They will eat nuts, almonds, kernels, chefnuts, figs, hemp-feed, hay-feeds, and a variety of fmall grains; they are fond too of blood, carrion, fat, fuet, and tallow.

In winter, if there be placed upon a window-fill, fome hemp-feed in a little box, without a lid, and fome fuet in a net, the Titmice will be attracted in confiderable numbers; their manner of taking the kernel out of the feed, which they place between their feet, and hammer with the bill, and the thievifh activity which they difcover in feizing this grateful plunder, is very amufing. The Titmice are a very prolific Genus; they lay from 18 to 20 eggs; fome of them make their nefts in holes of trees, employing their little beaks to give them a form fuitable to their purpofe, and others build theirs in an oval form, very large, and well adapted for fo great a number of eggs;

fome

some again suspend them ingeniously from the branch of a tree, and all employ a variety of proper materials to make a comfortable receptacle for their young ; and exert an amazing, and well directed activity, in furnishing their numerous offspring with convenient food, and in defending them from the assaults of every enemy.

No birds discover more fire and intrepidity ; they are the first to assault the owl ; they aim boldly at his eyes ; their attack is accompanied with a ruffling of the feathers, and a rapid succession of violent attitudes and precipitate motions, which express with energy their little rage.

When they are taken, they bite the hand and fingers of the bird-catcher, striking repeated blows with their bills, and calling with reiterated screams to their companions, who assemble in numbers, and joining in the cry bring numbers more.

They are easily taken in almost every kind of trap. Sometimes they are intoxicated by meal, or meat, soaked in wine ; they attempt to fly, find themselves giddy, flutter, fall, roll about, make an effort to rise, and fall again, and by their strange and singular gestures, and whimsical atti-tudes, contribute to the amusement of a young spectator.

Although Titmice occasionally discover a sa-
vage

vage difpofition, in general they live harmonioufly together; their cruelty feems capricious, and rather the confequence of their hafty, volatile character, than a fixed habit. Mr. Buffon faw one, who, fo far from abufing the power which he derived from his fuperior ftrength, when he might have done it with impunity, difcovered that tendernefs, which helplefs, and unprotected weaknefs fhould always excite in the ftrong.

Having put two young black-headed Titmice, juft taken from a neft, into a cage, with an old blue Titmoufe, it adopted the young, performed all the kind offices of a parent, dividing with them its food, and preparing it for them.

Some perfons pretend that at certain feafons they have a pleafing fong, and are capable of being taught to pipe tunes, but this is not a very general opinion, and may want confirmation.

Almoft all this Genus make hoards of provifions, but it arifes perhaps from the impulfe of avarice, rather than of forefight, fince they generally pafs the fummer in hilly fituations, and the winter in thofe that are more fheltered.

They feek retired fituations to rooft in; they feem to wifh to pierce the walls to make an evening retreat, and always at fome diftance from the ground; even in cages, they never rooft at the bottom.

Some species have been observed to pass the the night in holes of trees; they enter hastily, after looking about on all sides, as though to assure themselves that they were secure, and in vain has a stick been introduced, in order to disturb them, they have obstinately remained in their retreat.

It is probable that they generally return to the same situation to roost, for there they generally establish their magazines. In the spring they are very mischievous in gardens, by pecking the young buds off the trees, or by bruising them in search of insects.

The TOUPET TITMOUSE.

The head is crested, the forehead black; the body ash-coloured; beneath white, tinged with red.

It inhabits, breeds, and passes all the year in Carolina and Virginia. It confines itself to the woods, and like other Titmice feeds on insects. It flies swiftly, and during flight frequently folds up the wing, uttering at the same time a weak note.

The CRESTED TITMOUSE.

The top of the head is crested with long black feathers, margined with white; the chin and throat are black. It is white beneath.

This

GREAT HEADED TITMOUSE.

Published June 1.ˢᵗ 1787 by Jos. Johnson Sᵗ Pauls Church Yard London.

This beautiful bird exhales a natural perfume, which it contracts among the juniper trees, and other aromatic shrubs, that grow in the solitudes which it frequents. It delights in retirement; forests of pines, and thickets of juniper, are the retreats it loves; withdrawn from other birds, and far from the haunts of men, it enjoys in security that liberty, which seems so essential to its happiness! For, if it be once taken, which from its natural shyness, and distrust, is seldom the case, it sullenly refuses food; and whatever arts may be employed to soothe confinement, it resolutely refuses life, when deprived of liberty.

It inhabits Europe, and is found in several parts of France, particularly in Normandy, and in other latitudes between France and Sweden. It is as fruitful as other birds of the same genus.

THE GREAT TITMOUSE.

The bill, the head, and throat, are black; the cheeks are white; the back and wings, are of an olive green; beneath greenish; but an irregular stripe of black divides the whole of the stomach into two parts; there is a bar of white on each wing; and the legs are of a lead colour.

It inhabits Europe, is very frequent in England, and is found in many latitudes, between Sweden, and the Cape of Good Hope, in Africa.

H 5

This

This species frequents both hills and plains, bushes and groves, orchards and forests ; in general preferring elevated situations, particularly in summer.

In the spring, it is said to have a pleasing song, and some have professed to esteem it highly, as a call-bird, for the purpose of ensnaring others.

It is easily tamed, and like the Goldfinch may be taught to eat out of the hand, to draw water in a bucket, and even to breed in confinement.

In a wild state they pair early in February, but it is some time before they begin to build their nest, which they generally place in a hole of a wall of deserted houses, or in the hollow of a tree. They lay from eight to twelve eggs, and hatch them in twelve days.

They breed three times a year, and live to the age of five years.

THE BLUE TITMOUSE.

The crown of the head is blue, the forehead and cheeks, white ; a line of black extends to the back part of the neck, which is black. The back is of a yellowish green. Wing coverts blue ; the quills black, with dusky edges. The tail blue.

This is one of the commonest species of little birds ; they are easily taken in traps, and are remarkable for the colours of their plumage.

In

In our gardens they are confidered as very de-ftructive, not only injuring the young buds of trees, but taking off the fruit with fingular ad-drefs, which they carry to their magazine.

Fruit, however, is not the only food of the Blue Titmoufe, like other birds of the fame family; it feeds on infects and carrion, and upon the dead bodies of little birds : it takes off their flefh with fo much exactnefs, that it has been propofed to give them their little carcaffes to anatomife. This minute bird is remarkable for the courage, indeed the fury, with which it affaults the owl.

It does not always, (like moft of the Titmoufe genus), pierce hempfeed with the point of its bill, but fometimes breaks them between the man-dibles.

The female makes her neft in holes of walls, or trees, warmly lined with feathers, and lays from eight to twenty eggs. She is fuppofed to lay but once in a feafon, unlefs fhe has been induced to forfake her firft neft, and this fhe is very apt to do, if the eggs be touched ; but when once the young are hatched, fhe defends them with great courage.

In the winter they frequent the neighbourhood of houfes, with others of their genus, but with-out any appearance of union, or attachment.

H 6 Though

Though in fome refpects this little bird be mifchievous in our gardens, in others it is ferviceable.
We are indebted to it for deftroying many caterpillars, and the eggs and larvæ of feveral infects,
which are devourers of fruit.

The COLEMOUSE.

The head is black; the body afh-colour; the back part of
the head, and the breaft, white.

This is a very pretty bird; it is fmaller than the
Blue Titmoufe, and there is a neatnefs, and mo-
defty in the colour of the plumage, that produce
a pleafing effect.   It is lefs diftruftful than others
of its genus, and however often it may have been
caught, it feems to acquire no caution from experience, but again ventures into the fame fnares.

The Colemoufe is a courageous bird, and in all
its habits refembles the Blue Titmoufe; like that
running up trees, and along the branches, in every
direction, in fearch of infects.

The MARSH TITMOUSE.

The head is black; the back afh-coloured; the temples
white.

By fome, this bird is not thought a diftinct fpecies, but only a variety of the Colemoufe.

It prefers woods, rather than orchards and gardens,

LONG TAIL TITMOUSE.

_____ Published June 1st 1787. by Jos Johnson St Pauls Church Yard London.

dens, feeding upon feeds, wafps, bees, and caterpillars, and making a ·hoard of hemp-feed,
when it has an opportunity.

He collects feveral feeds in his bill, and carries
them to his magazine, to eat them at his leifure;
perhaps his manner of breaking the feeds, which
requires a fituation where he may place them
one at a time between his feet, and pierce them
with the beak, may compel him to this inftance
of forefight.

· The Marfh Titmoufe is a folitary bird, frequenting willows, and alder trees, and confequently marfhy fituations. It is common in England, and many parts of Europe.

The LONG TAIL TITMOUSE.

The head is white, the tail longer than the body, a broad
ftreak of black unites at the back part of the head, and paffes
down the back to the tail.

This Titmoufe has a fmall, and proportionably
a longifh body, which, together with the length of
its tail, as its flight is rapid, would induce one to
imagine that it was an arrow darting through
the air.

It inhabits the woods, is fprightly, and active,
never fcarcely a moment in repofe, flitting from
bufh to bufh, running along the branches in every
direction,

direction, hanging by its feet, or affembling at the cry of its own fpecies; it feeds on infects and feeds, pinches buds of trees, and lays from ten to feventeen eggs, like others of the fame family.

In the conftruction of its neft, the Long Tail Titmoufe differs from moft of the tribe; it is not made in the hole of a tree; this would be inconvenient to a bird of fo long a tail, the feathers of which are apt to fall off upon the flighteft violence; but it is firmly fixed upon the branches of a fhrub, about four feet from the ground. The neft is of an oval form, clofed above, with a fmall hole at the fide for an entrance, and fometimes another oppofite, that the bird may leave it without hazarding an injury to its tail, by turning round in fo fmall a compafs. The outfide is compofed of blades of grafs, mofs, and lichen; the infide well furnifhed with feathers.

They live together with their young in a family, during the winter. Their feathers are very long and downy. In the fpring they have a little fong.

This fpecies inhabits Europe, and even Jamaica, and is very common with us, frequenting gardens and orchards.

THE

The PENDULINE TITMOUSE.

The forehead is black; the head and neck afh-colour; behind the top of the head whitifh. There is a black band which paffes by each eye. The wing and tail feathers are dufky, margined with whitifh afh-colour.

This bird inhabits Ruffia, Poland, Italy, and fome other parts of Europe.

The great ingenuity which it employs in the conftruction of its neft, is the moft interefting circumftance in its hiftory. The outfide is formed of fibres, and little roots, but the bulk of it confifts of the down of willows, of the poplar, of cotton grafs, of thiftles, and of other plants : this is curioufly combined, and interwoven, until it acquires a texture almoft as firm as cloth; the infide is furnifhed with the fame materials, but not worked in the fame manner, that it may be foft and warm for the young. It is covered above, for the purpofe of warmth and fhelter, and fufpended from the forky part of a flender branch, by a twine, curioufly woven by this ingenious mechanic, of the filaments of nettles, and of hemp.

This neft is contrived to hang from a flender branch; hence it is gently rocked by the wind; and fufpended over a running water, which fecures it from the attempts of rats, fnakes, lizards,

and

and enemies of that kind ; befides, this fituation is remarkably favourable, on account of gnats, and infects, which are the principal food of the Penduline Titmoufe. The opening of the neft is on one fide, generally towards the water ; the female of this fpecies lays but four or five eggs, and breeds twice a year.

Penduline Titmice frequent marfhy fituations, concealing themfelves amongft rufhes, reeds, and the leaves of aquatic plants.

The BEARDED TITMOUSE.

The head is grey, the tail longer than the body, the bill orange colour in the living bird, beneath each eye is a tuft of black feathers, like whifkers.

It inhabits Europe, and is found in marfhy fituations, in feveral parts of England, in the marfhes among the reeds near London, and in Gloucefterfhire and Lancafhire.

This has been fuppofed to be an Indian bird. Mr. Buffon fuggefts, that it was introduced into England by the Countefs of Albemarle, who brought a cage full of them from Denmark, fome which might have efcaped, and planted a colony with us, but they are too numerous, and too widely difperfed, to fuppofe that to have been their firft introduction. They feed on the feeds of reeds, and on fmall
infects,

infects, and stay with us the whole year. Their nests are not very certainly known, but as some have been found composed of very soft, downy materials, and suspended from three reeds connected at the top, for that purpose, it is supposed that they belong to this species. The reed grounds in several places cover many acres, and are only visited in boats, at the time that the reeds are cut; this accounts for our knowing so little of the habits of this bird, which probably may be very interesting, if it be true, as asserted, that the male covers the female with his wings, when they repose. This singular attention might lead one to presume a number of other attentions, curious, and amusing.

THE AMOROUS TITMOUSE.

ts plumage is of a slate colour.

It inhabits the northern parts of Asia.

When birds of this species are confined in a cage, in pairs, nothing can exceed their fondness : they are perpetually billing and caressing one another, and by a thousand little attentions, not only alleviate the rigours of confinement, but make even captivity delightful.

THE

THE CAPE TITMOUSE

Inhabits the southern parts of Africa, and builds a nest in the form of a bottle, with a short neck.

It is composed chiefly of a cotton-like substance, and concealed amongst the thickest shrubs: on the outside of the nest, there is an additional structure for the lodgement of the male, whilst the female sits, or broods her young. It is said too, that when the female leaves the nest, the male strikes against the outside, and causes the edges of the entrance to close together; by this means, defending their young from such insects as might injure them, when unprotected by the presence of their parents.

GENUS 79. HIRUNDO.

The bill is short, broad at the base, small at the point, and a little bending. The gape of the bill very long.

The nostrils are open.

The tongue short, broad, and cloven.

The tail in most species forked, and the wings long.

The legs, the toes placed three before, and one behind, a very few species excepted.

This genus of birds lives on insects which they take on the wing.

Their

TITMOUSE.

Published June 1. 1787 by Jos Johnson S.t Pauls Church Yard London.

Their plumage, though not beautiful in itfelf, is gloffed with different fhades, fo as to fhew a change of colours, in different points of view.

The mouths of the Hirundo genus, or Swallows, are not like thofe of the Goatfucker, furnifhed with a vifcous, or clammy fubftance, to retain the flies and infects that they once have taken into their beaks : this is unneceffary to the Swallow tribe, whofe fight is remarkably quick ; they feize the flying infects by a fudden fnapping of the mandibles, which can be heard at a diftance.

Swallows are very fociable birds ; they come, and fly, and migrate in confiderable flocks, and in fome degree perform the kind offices of focial life, by affifting one another in the conftruction of their nefts.

Their nefts in general are formed with confiderable care, and intelligence. Some fpecies build in the holes of walls, or in holes which they themfelves prepare in fand rocks : they make the cavities fufficiently deep to infure the fafety of their infant brood, and carry them fuch materials, as may enable them to lie foft and warm, and at their eafe ; others again build againft houfes.

The flight of the Swallow is rapid, yet eafy, and continued ; and indeed, this feems its natural ftate ; it eats, drinks, and bathes, and even feeds its young upon the wing.

It

It rifes, defcends, and fails fmoothly, yet rapidly through the air, without the leaft appearance of effort.

It feels itfelf in its proper element, and as it glides through the yielding expanfe in every direction, by a cheerful twittering note, expreffes its felicity.

One time it purfues the flitting infects, following their oblique, and irregular direction, with the utmoft facility, quitting one to chace another, and in its flight feizing perhaps a third. Sometimes the Swallow fkims lightly over the furface of the fields, and of the water, to feize the infects, which the rain, or moifture have attracted; fometimes too it efcapes itfelf from the impetuous attempt of a bird of prey, by the ready quicknefs of its movements. Always mafter of its flight, however rapid, in an inftant it can change its direction. It feems to defcribe in the air, a changeable, tracklefs, labyrinth; the paths of which crofs, interweave, diverge, approach, confound, combine, rife, defcend, lofe themfelves, and appear again to interfect, and entangle one another in a thoufand ways, too complicated to be pictured to the eye, by the art of drawing, or to he imagination, by the powers of verbal defcription.

The

The Swallow tribe appears to belong equally to both continents. Indeed what country can we suppose inacceffible to birds who fly fo well, and tranfport themfelves from one place to another, with fuch wonderful facility.

With regard to the migration of Swallows, naturalifts are very much divided.

There are three opinions on this fubject.

The firft is, that they remove in the winter to warmer climates, in fearch of infect food.

The fecond, that they retire to caverns, and hollows of rocks, and pafs the winter there, torpid.

The third opinion is, that they conceal themfelves under water, in the winter, collecting (as fome fay), in numbers on a reed, until it break, and let them gently fink. Others relate, tha fe-veral of them take a ftraw between their beaks, and plunge toge-ther beneath the fu face; whilft others again affert, that they unite their feet together, and immerfe themfelves in clufters.

The advocates for the firft opinion, *that they remove to warmer climates,* quote feveral authorities.

Peter Martyr fays, that he *knows* that Kites and Swallows quit Europe at the approach of winter, and pafs into Egypt.

Father Kirker afferts, upon the teftimony of the inhabi-tants of the Morea, that a great number of Swallows and Storks pafs every year from Egypt and Lybia, to Europe.

Mr. Adanfon, an attentive obferver, and highly defcrving attention, afferts, that Chimney Swallows arrive at Senegal about the 9th of October, and return in the fpring. That on the 6th of October, being fifty leagues from the coaft, between the Ifland of Goree and Senegal, he faw four Swal-low.

lows, which rested on his ship; they were all so fatigued, that they suffered themselves to be taken, and he knew them to be at least of the European species.

About the same season, in the year 1765, the Viscount Querhoent, an intelligent Natural Historian, relates, that the Penthievre, a French ship, was almost darkened by a cloud of Swallows, near the Cape de Verd Islands.

Leguat, on the 12th of November, saw four Swallows, which accompanied his vessel to the Cape de Verd Islands.

Christopher Columbus, in his second voyage, saw one which flew near his ship, ten days before he discovered Dominica.

Forster, and many other travellers, and writers of respectable authority, speak decidedly to similar circumstances, of Swallows flying near ships, at some distance from land: and other attentive observers, have asserted, that *Swallows* leave England about the latter end of September; that prodigious swarms assemble at that time on the Coast of Suffolk and Norfolk; that they rest on trees, churches, and other buildings for some days, if the wind be unfavourable; but that if it should change in the night, and become propitious, they all disappear before morning.

Mr. Collinson, a very respectable Member of the Royal Society, mentions two curious relations, of good authority; the one communicated by Sir Charles Wager, the other by a Mr. Wright, a master of a ship. Sir Charles Wager writes, ‘ Returning home in the spring of the year, as I came into ‘ sounding, in our channel, a great flock of Swallows came, ‘ and settled on all my rigging; every rope was covered; ‘ they hung on one another like a swarm of bees; the decks, ‘ and carving, were filled with them: they seemed almost ‘ famished and spent, and were only feathers and bones, but

‘ being

' being recruited by a night's reft, they took their flight in
' the morning.'

Mr. White, an obfervant Naturalift, and the author of
an ingenious work, intitled, The Natural Hiftory and An-
tiquities of Selborne, on Michaelmas-day, 1768, early in
the morning, which was very mifty, on a large wild heath,
faw numberlefs Swallows cluftered on the bufhes; the moment the fun broke out, they were inftantly on the wing,
and proceeded with an eafy, and placid flight, towards the
fea. After this he faw no more flocks, only now and then a
ftraggler.—*See Philofophical Tranfactions, volume 51, part 2,
page 459.*

In Kalm's Voyage to America, when he had paffed over
about two-thirds of the Atlantic Ocean, on the fecond of
September, a Swallow fettled on his fhip.

This is, at leaft, a ftrong prefumption of their power to
perform a very diftant flight.

*In favour of the fecond opinion, that they retire to caverns,
and the hollows of rocks, and pafs the winter there, torpid,*

Several authorities are quoted. Ariftotle afferts, that many
have been found under thofe circumftances, without a fingle
feather upon their bodies. Albert, Auguftin, Nyphus,
Gafpard, Heldelin, and others, affert, that Swallows have
often been found in Germany, in a torpid ftate, in the hollows of trees, and even in their nefts.

Mr. Collinfon, in the 53d volume of the Philofophical
Tranfactions, page 101, mentions the teftimony of three
gentlemen, who fay, that a number of Sand Martins were
drawn out of a cliff, on the banks of the River Rhine, in
March

March 1762. And the Hon. Daines Barrington relates, on the authority of Lord Belhaven, that numbers of Swallows have been found in old dry walls, and sand hills, near his seat in East Lothian, in Scotland, not only once, but year after year; and that when exposed to the warmth of fire, they revived.

Some years ago, it is said, they were seen in a torpid state, on the fall of a great fragment of the chalky cliffs of Suffex; and in a decayed hollow tree, cut down near Dolgelli, in Merionethshire, in Wales.

In a cliff near Whitby, in Yorkshire, when digging for a fox, whole bushels of torpid swallows were found.

Mr. Conway, of Syckton, in Flintshire, in Wales, asserts, that a few years ago, looking down an old lead mine, in that county, he observed numbers of Swallows clinging to the timbers, seemingly asleep; that on throwing some gravel upon them, they just moved, but did not attempt to fly; this happened between the latter end of October and Christmas.

On the 23d of October, 1767, a Martin was seen in Southwark, flying in and out of a nest, and on the 29th of October, four or five Swallows were observed to hover about, and settle on the county hospital at Oxford; and once near Christmas, a few were noticed on the moulding of a window of Merton College.

The advocates for this opinion, support their theory by the analogy of Bats, of Marmots, of Dormice, and Bears, who pass the winter in a torpid state; and urge, that the prodigious exertions of this little bird, who, during the summer, has been so much on the wing, may require the refreshment of a winter's sleep. On the other hand, it is objected, that admitting these facts to be true, they may have been accidental circumstances. The birds so found, might have been a late hatch that were left behind, not being sufficiently ad-

vanced

vanced to support the fatigue of a migration; or at most, that these facts may apply only to some species. Nor should it be omitted, that in 1757, by the direction of Mr. Collinson, a bank of a river, perforated with the holes of the Sand Martin, was accurately searched, in order to find some of these birds, but without success. It is farther objected, that if this opinion, of their torpor arising from cold, was well founded, when the seasons are mild, in autumn, they ought not to disappear so soon; and when we have warm weather in February or March, they should appear sooner; this, it is said, is not the case.

The third notion, that they remain concealed under water, or in the mud, during the winter, is still more extraordinary; and from its being contrary to all analogy, seems to require the strongest support from actual observation.

It was first suggested by Olaus Magnus, Archbishop of Upsal, who says, that they are often found in clustered masses at the bottom of the northern lakes, beak to beak, wing to wing, and foot to foot. That when they are taken by experienced fishermen, they throw them in again; but those who are unexperienced, by exposing them to warmth, revive them, it is true, but that this premature resurrection is attended only by a short renewal of their powers.

Some very respectable Natural Historians have favoured this opinion; amongst others, the father, and founder of this science, the great Linnæus. Klein has adopted the same opinion, and even produced certificates; they are principally signed by one person, and speak of a solitary instance, which happened long before, in the years of childhood, or else are founded upon hearsay, and admitted to be uncommon.

PART VI.   I   Etmulles

Etmulles fays, that he faw a bufhel of Swallows, cluftered together, taken out of a frozen fifh pond.

Dr. Colas, fpeaking of the manner of fifhing in northern climates, by breaking holes, and drawing nets under the ice, afferts, that he faw fixteen Swallows fo drawn out of the Lake of Samrodt; thirty out of a great pond belonging to Rofincilen; and at Schlebitten, near a houfe of the Earl of Dolma, he faw two Swallows juft come out of the water, that could fcarcely ftand, being very wet and weak, and with the wings hanging on the ground.

To fupport thefe furprifing, and confeffedly rare inftances, by fome kind of analogy, they refer to the cafe of the larvæ of many infects, of frogs, the amphibious animals, and fifh.

It is objected, that Olaus Magnus, who broached this doctrine, feems to have been very credulous, or very fortunate, that at the fame time that he has peopled the water with Swallows, he has tranflated mice to the clouds; and that his writings are very amufing, and abound with wonders.

That Linnæus reftrains his affertion to two of the fpecies; that he does not profefs to have feen, or to vouch the truth of every thing he relates.

On the authority of Keoping, he has rather favoured the opinion, that there are men with tails.

On Solander's authority, he has given a wonderful account of the Furia, an infect in Bothnia, which falling out of the air upon the bodies of men, or animals, penetrates them in an inftant, and afflicts them with pain fo excruciating, as to prove fatal in a quarter of an hour.

On the authority of Baron Munchaufen, he fpeaks of the feeds of Fungufles, which being difperfed in water, live, and move, at laft fix themfelves, and become Fungufles again.— Animals thus becoming vegetables.

It

It is further objected that the inflances adduced are very few, and moll of them grounded upon hearfay.—That if this really were the fact, they would often be taken in nets by fifhermen, and the proofs be notorious, from the frequent experience and teflimony of failors, travellers, fowlers, and ruflics. That as their collecting together is a circumfkance of general obfervation, fo their actual immerfion, and emerfion, muft have been frequently feen, if it really exifted.

It has been publickly advertized in Germany, to any perfon, who in the winter fhould produce Swallows, taken out of water, in this torpid flate, to recompenfe them, by paying them the weight of the Swallows fo found in money.

Many literary charafters, and perfons of diflinction, who were difpofed to believe this circumflance, have promifed to endeavour to furnifh additional proofs to Mr. Reaumur, and Count Buffon, but have not done it.

It is contended, that the analogy between Swallows, and amphibious animals, does not flrictly apply, as the laft are anatomically different; that having only one auricle, and one ventricle, and cold blood, their ftructure is calculated to admit of their breathing arbitrarily, or at diftant intervals, whereas there is no fuch contrivance in Swallows. And that very ingenious anatomift, Mr. John Hunter, has diffected many Swallows, but found nothing in their organs of refpiration different from other birds. It is contended too, that lizards and frogs, which do fleep during winter, do breathe in their torpid flate, confequently that the notion that terreftrial animals can remain long under water, without drowning, feems unfounded on obfervation, and improbable.

The authors of the Italian Ornithology, and Count Buffon, have made the experiment of plunging Swallows under water, and they died.

I 2

Count

Count Buffon tried a fimilar experiment with frogs and fifh, in the month of February; thofe which he put into water, which was open to the air, and were allowed to rife to the furface, continued to live ; but thofe that were put into a veffel of water, under circumftances exactly fimilar, except that they were reftrained from coming up quite to the furface, foon fhewed fymptoms of uneafinefs, and died ; fome in fix hours, and thofe that furvived the longeft, in two days.

Mr. Frifch tied to the legs of Swallows, threads dipped in water colours ; the next year, he afferts, that he faw the fame birds, and that the threads had not loft their colour.

ADDITIONAL OBSERVATION.

Mr. White, the ingenious author of the Natural Hiftory of Selborne, whofe remarks well deferve the attention of every ornithologift, from repeated obfervations, inclines to the opinion, that many of the Swallow kind do not depart from this ifland, but retire to holes and caverns, from which they are allured even in the winter, when the weather is warm and inviting.

THE CHIMNEY SWALLOW.

The forehead and chin, a reddifh chefnut colour ; the plumage above blackifh, gloffed with purple ; beneath white ; the tail forked ; all the feathers in the tail, except the two in the middle, marked with an oval white fpot, near the end.

The Chimney Swallows feem by inftinct attached to the habitations of men. They build their nefts in our chimnies, and even in chambers,

bers, but little frequented, if the windows be left open.

In mountainous countries, where the chimnies are clofed above in the fummer, on account of the heavy fnows, they build under the eaves, ftill conftant in their attachment to the neighbourhood of man. A bewildered traveller when he fees one of thefe birds, may confider it as a bird of good omen, prefaging that fome habitation is near.

They generally appear about the beginning or middle of April. Their vifit is not haftened by premature warm weather, in February or March; or retarded by fevere cold in April. In 1740, great numbers died in France: the feverity of the cold had deprived them of their infect food; their bodies were emaciated, and fome were obferved to fix themfelves to the wall, and feize the dried infects that remained in the fpiders webs.

Birds fo innocent, and ufeful, have every claim to our protection: they deliver us from thofe fwarms of infects which would otherwife infeft our houfes, injure our gardens, our trees, and, perhaps, our harvefts. The leaft return we ought to make, and, indeed, the only one they require, is, to let them live fecure. Yet, we fee, that many, infenfible to the advantages we receive from them, indulge themfelves in the inhuman,

I 3   and

and unmanly amufement, of fhooting at them with guns, under the pretence of perfecting themfelves in that art, by exercifing it againft a mark, which moves fo fwiftly, fo irregularly, and confequently which it muft be fo difficult to hit.

They frequent the fame place, year after year; this is proved by the experiment of Frifch : and at a caftle near Lorraine, a ring of brafs wire was fixed to the foot of a Swallow, the Swallow returned with it again the next year. Every feafon they build a new neft, clofe to that of the year before, compofed of ftraw, of plaiftered earth, and horfe-hair, and lined with dry grafs and feathers, and open at the top.

They are careffing birds; the female lays twice a year. Whilft fhe fits, the male fpends the night on the edge of the neft ; his reft is fhort ; he flies till late in the evening, and begins his babbling early in the morning. As foon as the young are hatched, both the parents frequently feed them, and keep the neft remarkably neat. But it is very amufing to fee them inftruct their offspring how to fly ; they encourage them by their voice, prefent them with their food at a little diftance,, drawing back by degrees as they advance, and at laft gently pufh them from the neft, not without the appearance of great folicitude : they fport with

their

their young ones as they fly, and warble to them in the moſt expreſſive manner, as it were to animate their endeavours, by the aſſurance of protection.

As an inſtance of their ardent attachment to their young, Boerhaave relates, that a Swallow that had been abſent to get proviſions, at his return finding the houſe on fire, to which his neſt was fixed, darted through the flame, to feed, and protect his offspring.

Though Swallows paſs much of their time upon the wing, yet they often repoſe themſelves upon the roofs of houſes, chimnies, trees, and ſometimes on the ground.

In England, towards the cloſe of ſummer, they have been obſerved to paſs the night upon alders, and aquatic trees, or ſhrubs; for this purpoſe they chooſe the loweſt branches that are moſt ſheltered from the wind.

Two gentlemen, who ſpent a night at Maidenhead-bridge in September, went by torch-light to an adjacent iſle, and in half an hour brought away 50 dozen. They had only to draw the willow twigs through their hands, the birds did not attempt to eſcape: if it be ſaid that theſe birds were aſſembled in this manner, in order to plunge under the water, it will be objected, that in that

I 4   caſe,

cafe, in a river fo perpetually fifhed as the River Thames, the fifhermen muft very frequently bring them up in their nets.

It has been obferved, that the twigs upon which they have collected, generally die.

They leave this country, or at leaft difappear about the beginning of October; and it is fup-pofed they generally depart in the night, per-haps to be lefs expofed to the attacks of birds of prey; and that they avail themfelves of a fa-vourable wind, which muft wonderfully facilitate their paffage: when we confider the velocity with which air balloons have been carried, mere-ly by the current of the air, at a little elevation, the difficulty of the migration of birds feems much leffened; and it has been obferved by Mr. Hebert, that Swallows, on their departure, rife into a higher region of the air.

In the Iflands of Hieres, which are on the fouth coaft of France, where the weather is always tem-perate, and a perpetual fpring is enjoyed, Swal-lows have been feen all the winter; there they have found their infect food. They rooft upon the orange trees, and injure that delicate plant.

A Swallow has been employed like a Carrier Pigeon; the female was taken from her neft, to the place from which the intelligence was to be fent, and a thread of a certain colour, and a cer-

tain

tain number of knots, was tied to the foot; the affectionate mother haftened to her neft, and brought with aftonifhing expedition, the account which was confided to her.

In England we call this bird the Chimney Swallow, becaufe it generally builds in chimnies, fometimes at the depth of five or fix feet, fo that the young find a difficulty in leaving it, and frequently fall into the room below. When they have fucceeded, they remain for fome time on the chimney top. Their next effort is to reach the leaflefs bough of a tree, where they fit in a row; foon after they begin to fly, their parents attend upon, and meet them occafionally at an angle, with an infect which they receive upon the wing.

In Sweden, they build in barns; in other countries, in porches, gateways, galleries, and open halls.

They lay from four to fix eggs; their firft brood leaves their neft the firft week in July, their laft about the middle or end of Auguft.

The following remarks on the Chimney Swallow, are extracted from Mr. White's interefting publication, intitled, The Natural Hiftory of Selborne.

On the 22d September, 1772, he obferved the

I 5   Swallows

Swallows and Martins collecting on a walnut-tree, the next morning, which was foggy, at the dawn of day, they all arofe together; the rushing noife of the wings, of fuch a fwarm of birds, againft the hazy air, was heard to a confiderable diftance; after that day, only a few ftragglers were feen.

He has remarked in fome late fprings, that though they made their appearance about the middle of April, yet meeting with cold, bluftering, north-eaft winds, they immediately withdrew, abfconding for feveral days, until warmer weather allured them out; hence, he infers, that they do not migrate.

When a boy, he remembers to have feen a ftraggler on Shrove Tuefday, which muft have been not later than the middle of March, and often happens early in February. He has known a Chimney Swallow build in the fhaft of a well; but he adds, that they generally with us build in chimnies, preferring a funnel contiguous to thofe where there are conftant fires, and difregarding fmoke.

He remarks, that the Swallow difcovers wonderful addrefs in afcending and defcending through fo narrow a paffage. When hovering over the top of the funnel, the vibrations of her wings in the confined air, make a rumbling noife like thunder.

The

The Swallow, he conceives, chufes this, in fome refpects inconvenient fituation, to fecure the brood from rapacious birds, particularly owls, which frequently fall down chimnies, probably in attempting to get at the neftlings.

When the young Swallows can fly, but are ftill incapable of providing themfelves with food, they play about whilft their parents are chacing flies, who, when they have collected a fufficient quantity, make a fignal, on which, the parent and the young, rifing towards each other, meet at an angle, and the young one receives the food, uttering a little note, expreffive of gratitude, and affection. They fip water as they fly, and bathe upon the wing.

They attend horfemen for miles together, as they ride over the downs, fporting before, behind, and wheeling round, and collecting the infects which are difturbed by the trampling of the horfes.

They feed much on fmall coleopterous, or fheath winged infects, and fettle on the ground, picking up gravel to digeft their food.

The Swallow is a pleafing fongfter, and fings in foft, funny weather, between April and September. A Swallow for two years together built its neft on the handles of a pair of garden fhears.

Another built its neft on the wings and body

of an owl, that was hung from the rafter of a
barn. On placing a conch fhell the following
year, where the dead owl had hung before, a
Swallow built in the conch fhell, and both thefe
nefts were lately in Sir Afhton Lever's Mufeum.

The ESCULENT SWALLOW.

All the tail feathers are marked with white at the points.

The moft curious circumftance that we are ac-
quainted with refpecting this bird, is its neft. Mr.
Poivre gives the following relation: That in the
Streights of Sunda, near Java, he went on fhore
with a failor, on an ifland, called the Little Toe;
that in a deep cavern in the rocks, upon the beach,
they found a cloud of little birds, which in endeavouring to efcape, as they entered, darkened the
opening of the cavern; he beat down fome with
his cane. As he advanced, he perceived the roof
covered with nefts, in the form of cups, flattened
on one fide, containing each two or three eggs,
or young birds, and lined with feathers; he
brought feveral away with him, and made a drawing of a neft and the birds; he defcribes them as
very fmall, not exceeding in fize a Wren, or
Humming Bird.

The nefts were immediately difcovered to be of
that kind, which the Chinefe confider as fo great a
delicacy; ufing them in foups, and ragouts made
of chickens, and mixed with ginfeng.

The nefts are firft foftened by being fteeped in
water, then pulled to pieces, put into the body of
a fowl, and ftewed.

Mr. Poivre fays, that in the months of March
and April, the furface of the fea from Java to
Cochin-China, and from Sumatra to Guinea, is
covered with a vifcous fubftance, (perhaps fifh
fpawn) which refembles glue, or ifinglafs diffolved.
The Efculent Swallow is fuppofed to make its neft
of this matter, which it may collect as it flies, or
take from the rocks, where it may have been left
by the waves.

Mr. Poivre collected, and dried fome of this
fubftance, and found it refemble very nearly the
fubftance of the neft.

The colour of the new nefts is white, like
ifinglafs; and as the materials of which it is made
are foluble in water, the birds very wifely build
them in caverns, fheltered from rain.

Some of the nefts are of a darker colour; thefe
are fuppofed to be older, and mixed with impuri-
ties, and are ufed as glue.

The nefts weigh about half an ounce each; they

5   are

are an article of commerce, and, it is faid, that the Dutch export every year from Batavia 125,000lb. or between 50 and 60 tons weight of thefe nefts.

Some writers fuppofe, that they are com-pofed of fea worms, of the Mollufca tribe ; fome of a kind of cuttle fifh, and others of a glutinous plant, called Agal Agal.

Mr. Marfden fays, that the Swallows that make thefe nefts, are about the fize of a Martin, and that they collect with their bills from the foam of the fea, the materials with which they make their nefts.

THE MARTIN.

The tail feathers are without fpots. The upper part of the body, wings, and tail, blackifh, gloffed with blue ; white beneath ; and the extreme part of the body at the fetting on of the tail, white.

The Martin builds its neft againft the fides of cliffs that overhang the fea, but more frequently under the eaves, under cornices, in the upper cor-ners of windows, or under any other projecting parts of houfes. The materials that it employs are very fimilar to thofe ufed by the Swallows, viz. earth on the outfide, particularly that which has been thrown up by worms, tempered, and mixed with ftraw ; and lined with feathers. There is, however,

ever, this difference, that the Martin's neft is covered above, and the entrance is at a fmall hole
in the fide ; add to this, that the Martin's neft is
far from neat, it is often infefted with maggots,
fleas, and even bugs, and thefe are likewife found
in the feathers of the birds themfelves.

When they build their nefts againft rocks, it
is thought that they laft but one feafon ; but the
nefts which they fix to our houfes, ferve for feveral years in fucceffion.

Sometimes they complete their neft in five or
fix days ; for this purpofe they bring the earth in
their beaks and claws, tempering it, and placing
it with their bills ; fometimes feveral affift in the
conftruction of the fame neft. There have been
inftances too, where fome have been feen active
in endeavouring to deftroy the labours of others,
and to pull down the neft as faft as thefe built
it up.

Martins do not appear in England quite fo foon
as the Swallow, and for a little while after their firft
arrival, confine themfelves to low, and marfhy
fituations. They feem fo much under the influence of climate, and to have fuch a prefentiment
of approaching changes in the weather, that they
have been known to quit Lapland the beginning
of Auguft, abandoning their young, though the
weather

weather was then fine ; but it foon changed, and
by the 9th of September, fledges were ufed.

At other time they have been known to ftay
later, though the weather has not been very mild ;
it was judged from thence that no very fevere
weather would foon happen.

They breed two or three times in the feafon :
the male affiduoufly attends upon the female whilft
fhe fits, and fhews the tendereft folicitude for her,
and for their young, attacking with great fpirit any
bird that approaches the neft. Still there have been
inftances where, in confequence of an accidental
derangement, this affection feems to have va-
nifhed. One of their little ones, juft capable of
flying, fell from the neft upon the window fill ;
the parents entirely neglected him : finding himfelf thus abandoned, he exerted his powers, and
in three quarters of an hour began to fly. A neft
was taken with all the young from the upper corner of a window, and laid upon the window fill ;
the parents flying backwards and forwards, and
frequently vifiting the fpot where their neft was,
muft have feen their young, and heard their cries,
but they fhewed them no attention.

They live on infects, and feed their young when
they can fly, on the wing ; they catch the infects
flying, and if they fee one on a wall, they brufh

it

it with their wing, to induce it to attempt a flight, and thus the infect becomes more eafily their prey.

It is afferted by fome, that Martins alfo feed upon Caterpillars, which they pick from trees.

It has been obferved, and Linnæus has given his authority to the obfervation, that Sparrows fometimes diflodge the Martins from their neft, who, in revenge, plaifter up the entrance, and thus bury the invader alive.

Mr. Bomare has made a very entertaining ftory of this circumftance in his dictionary: he reprefents the injured Martins as gently expoftulating at firft with the intruders, but without effect; that then they call a council of the neighbouring Martins, ftate their grievance, and implore their aid; that the Martins depute heralds to fummon the invaders to retire, and that when they find thefe meffengers return without fuccefs, the whole affembly breaks up. Every Martin provides himfelf with fome tempered earth; they all refort to the neft, making a third attempt to induce the Sparrows to reftore the habitation which they have ufurped: this proving ineffectual, they proceed to punifh fuch obftinate injuftice by clofing up the entrance, and entombing the Sparrows, thus making the property they have fo unfairly acquired the means of their punifhment.

It

It muſt be ſuppoſed that there is a good deal of imagination in this detail of Mr. Bomare's ; indeed, Mr. Buffon doubts if Martins ever have recourſe to ſuch a revenge, having ſeen Sparrows uſurp their neſts, and obſtinately retain poſſeſſion, though perſecuted for ſome days by the Martins, who, however, made no attempt to cloſe the entrance of the neſt.

It would be difficult to bring up Martins or Swallows in confinement, becauſe inſects ſeem their proper food. There is an account of ſome children keeping a neſt of Chimney Swallows alive ten days, by feeding them with that which had paſſed through other Swallows before ; they lived very well upon this food, until the proceſs was interrupted by the mother of the children, who ſeemed to be more attached to neatneſs than to philoſophical experiments. Leguat, a traveller, ſpeaks of a Swallow that he tamed, and brought from the Canary Iſlands ; he ſuffered it to fly out in the morning, and it conſtantly returned in the evening.

The Count Buffon mentions a tame Swallow, or Martin, which had conceived ſuch an attachment to its miſtreſs, as to remain whole days upon her knees, and ſhew many expreſſions of joy on her return, after a ſhort abſence. It began to

feed

feed out of her hands. One day the bird made its escape, it did not fly far, but suffered itself to be caught by a child, and soon after fell a prey to a cat.

They leave England about the latter end of September, or beginning of October, and before their departure have been obferved to exercife themfelves in flying more than ufually high, as though preparing for a migration.

In the Natural Hiftory of Selborne, a book of great merit, lately publifhed, the author fays, that on the 26th of November, his neighbour faw a Martin in a fheltered bottom chacing flies, the fun then fhone warm; that on the 4th of November he faw feveral Houfe Martins playing all day long over his fields. He obferves, that from the fituation of the Martin's neft againft a perpendicular wall, it requires great effort, and judgment, firmly to fix the foundation that the fuperftructure may be fecure. For this purpofe the bird clings with its claws, and ufes its tail for a fupport like a Woodpecker; and that the materials may not fall by their weight, in a foft ftate, the intelligent architect advances its work flowly, building only in the morning, and devoting the reft of the day to amufement, and the chace of infects.

The parent birds remove every thing offenfive from

from the neft, whilft their brood are very young. The little ones foon acquire an attention to neat-nefs themfelves, and perform that office by putting their tails out at the opening of the neft.

Martins prefer a north-eaft or north-weft expo-fure for their nefts, that they may not be cracked, and injured by the heat of the fun; yet, inftinct does not feem to lead them invariably right, for fometimes they build in fuch fituations that their nefts have been wafhed down by rain, and yet they perfevere year after year in the fame fitu-ations.

Mr. White has feen young Martins that only left their nefts on the 22d of October: from this, and other obfervations, particularly from having traced their evening retirement in October, for feveral nights fucceffively, amongft fome low and ftunted fhrubs, almoft impervious to a dog, and in many parts difficult to approach, he fuppofes, that part of them, at leaft, muft have a winter retreat in this ifland. It is very remarkable too, that the numbers which return in the fpring, bear little proportion to thofe which withdraw the pre-ceding winter. Are they deftroyed by their ene-mies, or do the parents prevent the younger brood from fettling in their neighbourhood?

Martins

Martins have not much note ; they chirp, however, from May to September.

They are diftinguifhed from other birds of this
genus, by their legs being covered with foft,
downy feathers, to the toes.

The SAND MARTIN.

Is of a moufe colour ; the throat and ftomach white ; the
neck encircled with a moufe coloured ring.

It inhabits moft parts of Europe, and frequents
the fteep banks of rivers, of fand pits, of fand rocks
near the fea, and by the fide of ftanding waters,
becaufe their infect food abounds in fuch fituations, and becaufe there they can with greater
eafe, make thofe ferpentine, deep, but horizontal
holes in which they depofit their eggs, and hatch
their young.   The neft of this fpecies is only a
heap of ftraw and dried grafs, furnifhed properly
with feathers.

They have more abundant means of fubfiftence
than moft of their tribe, as they not only feed
upon infects which they chace with great rapidity
and addrefs, as they hover over the furface of the
water, but they feed upon the larvæ and chryfalids
which conceal themfelves in the ground, and their
young in confequence, in general, are very fat,
and

and in fome countries efteemed a great delicacy.
They arrive in England, and migrate from it
about the fame time as the Martin.   They have
been feen in fome parts of France during the winter months, but in very fmall numbers, fo as not
to juftify any general inference ; and as they have
refources in refpect of food which the Martins
have not, as mentioned before, it is eafy to conceive the probability of fome accidental ftraggler
furviving the winter.

Mr. White mentions, that they build in fcaffold
holes, in a ftable at Bifhop-Waltham, but that
this wall is in a retired, and fequeftered inclofure,
and faces a large and beautiful lake.   He makes
the following obfervations.

Many holes, of different depths, made by the
Sand Martin, are found unfinifhed at the end of
fummer: perhaps the birds in thofe places may
have met with ftrata too hard and folid for their
tender bills, or the foil may have been loofe, and
mouldering, and have fallen in, for it is fcarcely
conceived that thefe are provided for the fucceeding
year.   After fome years they forfake their former
holes, which may have become foul and offenfive
from long ufe ; and perhaps untenable, from their
abounding with fleas, which have been feen fwarm-
ing at the mouths of their holes, like bees on the
ftools of their hives.

The

The Sand Martins difcover great dexterity in boring their holes in the fand, and perform the tafk with furprifing quicknefs, confidering the feeble inftruments they employ ; indeed it is wonderful how fo weak, and tender a bill, can perforate the bank at all.

They have a peculiar manner of flying, flitting about with odd jerks, like the motions of a butterfly.

They are the fmalleft of the genus Hirundo, except, perhaps, the Efculent Swallow.

THE PURPLE SWALLOW.

The plumage is of a beautiful violet colour, and the tail forked.

It inhabits Carolina and Virginia, appearing in May, and retiring on the approach of winter.

The common people are very fond of this fpecies, and place boards and other conveniences for them to build upon, as fome do for Sparrows in England. Thefe little birds requite their kindnefs, by delivering them from flies and infects, and by alarming their poultry upon the approach of Hawks, and other birds of prey, which they attack with great fpirit.

THE

The SWIFT.

The plumage a dufky black, the throat white, the four toes all forwards.

This is a large fpecies ; it comes the lateft, appearing about the latter end of April, and departs the earlieft ; and during its fhort fummer refidence with us, has only time to rear one brood.

Its manner of flying is more rapid, and more elevated than that of others of the Swallow tribe ; and from the great length of its wings, and fhortnefs of its legs, if it were once on the ground, it would have great difficulty in recovering its flight; when this happens, it waddles in an aukward, embarraffed manner to fome elevation, a little hillock, or a ftone, from which with great effort it contrives to rife. From this ftructure in the bird, the whole of its life is fpent in the oppofite extremes, of the moft rapid motion, and the moft abfolute reft. Sometimes, indeed, it is feen to fix itfelf againft the wall, or the trunk of the tree near its neft, and to clamber into it with the help of its beak and wings, and tail and claws, availing itfelf of every fupport in its power ; but far more frequently it enters its neft on the wing, and with that rapidity that it feems in an inftant to vanifh nlſh from the eye, as though it were diſſolved in air.

Their neſts are generally made in holes of walls, which are larger within than at the entrance: they prefer elevated ſituations, as ſteeples, and lofty towers, though ſometimes, for concealment, perhaps, and ſecurity, they build under the arches of bridges ; ſometimes too, in hollow trees. They have been known to uſurp the neſts of Sparrows, and when Sparrows have intruded into theirs, they have contrived means of compelling them to relinquiſh the unjuſt acquiſition. Their neſts in towns are compoſed of various materials, of ſtraw, of graſs, moſs, hemp, thread, ſilk, rags, gauze, muſlin, and the ſweepings of the ſtreets ; ſome of theſe they take upon the wing as they are raiſed from the ground by the wind ; ſome, perhaps, they procure from Sparrows neſts, which they have been obſerved to plunder.

When the young are hatched, their parents feed them only two or three times a day, but then they bring them a plentiful proviſion, their ſwallow being filled with flies, beetles, butterflies, and inſects of various kinds ; they feed too upon ſpiders.

Both the young and the parent birds ſwarm with fleas.

PART VI.   K   ſm

In Savoy, the young are efteemed delicate food.

In the Ifland of Zante, children often take them by fufpending a feather upon a hook, at the end of a long thread from the window of a tower, or fome elevated building; the Swift feizing the feather to convey to his neft, is caught upon the hook.

Swifts are impatient of heat, and on that account pafs the middle of the day in their nefts.

Like others of their tribe, they are liable to be infefted with infects.

We are indebted to Mr. White, for the following obfervations on the Swift.

That it eats, drinks, collects materials for its neft, and performs every office on the wing, except fleeping, and incubation; he thinks he has frequently feen them carefs one another in the air.

The Swift never feems fo much to enjoy itfelf as in fultry weather, juft previous to thunder. In warm, funny mornings, in little parties, they wing their rapid flight round fteeples and churches, fqueaking in a clamorous manner: thefe are fuppofed to be the males ferenading their females as they fit, fince they feldom alter their cry until they approach the neft, and thofe within return a little note, expreffive of complacency.

2   They

They collect infects in a pouch under their tongue, and fome chace in higher regions of the air than others of their tribe ; though they have been obferved flitting rapidly very low for hours together, over pools and ftreams, in purfuit of the phryganæ, ephemeræ, and libellulæ (infects,) juft emerged from their larva ftate.

They are on the wing all day long, in gentle rains, whence we may infer, that their feathers are capable of refifting much wet. In windy weather, or heavy fhowers, they confine themfelves to their nefts.

When they arrive in the fpring, their plumage is of a glofly foot colour, except the chin, which is white ; but before they leave us, by being fo continually in the fun, it becomes weatherbeaten, and faded. They return again glofly the next feafon ; hence, it is probable, that they do not retire to funny climates, and this feems fupported, by their difappearing from the fouthern parts of Spain, before they become invifible with us. Do they withdraw to moult, and to reft from the fatigue of fome months paffed in unceafing activity ?

They breed, as obferved before, but once in a feafon, and produce only two young ones.

They are fearlefs whilft haunting their nefting

places : then they are not fcared by a gun, and are often beaten down with poles, and fticks, as they attempt to enter their nefts.

They have a ftrong grafp with their feet, which enables them to cling to walls, and their bodies being flat, they can enter a narrow crevice.

They are much infefted with vermin ; and young ones are fometimes found fallen on the ground under their nefts, which the fleas perhaps may have rendered infupportable.

For feveral years eight pair were obferved to frequent the Church at Selborne : as they every year bred eight pair more, what becomes of the increafe ?—the parents, perhaps, compel them to find fome other fpot.

On the 24th of Auguft, Mr. White obferved a folitary Swift, which he difcovered attended upon two young ones in a neft, under the eaves of a building:

On the 27th they all difappeared. On the 31ft he had the eaves uncovered, and found in a neft two dead Swifts, quite putrid, over thefe a fecond neft had been formed.

This proves that Swifts, when fuch a circum-ftance makes it neceffary, can fubfift here after the ufual time of their migration ; it likewife affords a prefumption, that they raife but one brood in a year.

THE

The WHITE BELLIED SWIFT.

The plumage is dusky; the throat and stomach whit : the four toes placed forwards, and it has ten feathers in the tail.

It inhabits the mountainous parts of Spain, Switzerland, and Savoy, building in holes of rocks. It appears in Savoy the beginning of April, frequenting ponds and marshes at first, for 15 or 20 days; after which it retires to the mountainous parts to breed.

The WHITE COLLARED SWALLOW

Makes its nest in houses at Cayenne.

The nest is large, composed of the down of dog's bane, well woven; the cavity is divided obliquely by a partition, and a small parcel of soft down is placed over the top, to keep the eggs and the young brood warm.

GENUS 80. CAPRIMULGUS.

The beak a little hooked, very small, depressed at the base; and beset with stiff bristles on the edges of the upper mandibles.

The gape is very wide.

The tongue is pointed, and entire at the end.

The tail consists of ten feathers, and is not forked.

K 3

The

The legs are short; the toes united by a membrane as far as the first joint; the claw of the middle toe in some species serrated.

The birds of this genus bear the same relation to Swallows, as Owls to the rapacious birds: their habits and manners are very similar, but like Owls, they take their food in an evening.

The Goatsucker feeds principally upon gnats, beetles, and moths. He begins his chace when the sun is near the horizon; and if he appear in the middle of the day, it is only when the weather is cloudy, his flight then is low, and not long continued, because his eyes are unpleasantly affected with the glare of light.

Like the Swallow tribe, he is not under the necessity of snapping his beak every time he gets an insect into his mouth, for it is furnished with a kind of glue, which entangles, and retains his prey.

Goatsuckers with us are migratory birds, arriving in England in May, and departing in August.

They do not give themselves the trouble of building a nest; a little hole upon the ground, in a stony place, at the foot of a tree, or of a rock, answers their purpose; there the female lays two eggs; she sits upon them very assiduously;

and

and it is faid, if fhe apprehend tl.ey are difcovered, that fhe very dexteroufly contrives to
pufh them into another hole, which, though as
expofed as before, may appear to her more fecure.

'The birds of this genus have a remarkable habit of flying round a tree, in full leaf, perhaps a
hundred times fucceffively, in a rapid and irregular manner, fometimes darting inftantly down, as
it were to feize their prey; at other times rifing as
inftantaneoufly, probably in chace of thofe infects
which are fluttering there.

As the Goatfuckers wing their flight with great
rapidity, and with their beaks open, they make a
whirring noife, which has been compared to that
of a fpinning wheel, from which they have been
called Wheel-birds.

They feldom perch, but when they do, it is
lengthways on the branch.

The EUROPEAN GOATSUCKER.

The plumage is very fingular, and not eafily defcribed;
the ground colour is almoft black, but it is mottled like the
Woodcock's.

The legs are fhort, and feathered below the Knee.
The noftrils are faintly tubular.

This is the only fpecies that is found in Europe.

K 4   It

It inhabits equally Europe and America; flies in the evening, feeding on moths, and nocturnal insects, particularly the Scarabeus Melolentha, or Dor Beetle. It is migratory, stays only from the end of May to the end of August in the northern, and until the end of September in the southern parts of the island.

It has two notes, the one a sharp squeak, this it utters as it pursues its mate; the other resembles as mentioned before, the humming of a spinning wheel. It not only makes this noise as it flies, but often begins it in the evening, sitting on a bare bough, with its head lower than the tail, and quivering the lower mandible. The noise jarrs so much, as to occasion a sensible vibration on any litt'e building, on which the bird may chance to alight.

Mr. White observed a Goatsucker playing round a large oak, which swarmed with Scarabæi Solstitiales, or Fern Chaffers; its quick evolutions, and powers of wing, exceeding, if possible, those of the Swallow genus. He·saw it distinctly more than once put out its short leg whilst flying, and by bending its neck, deliver something into its mouth, probably some of the Chaffers which it might catch with its feet: this, Mr. White thinks, may justly account for the
middle

middle toe being furnifhed with a ferrated claw, to retain its prey.

The VIRGINIA GOATSUCKER

Arrives in Virginia about the middle of April. A few minutes after the fun is fet, they frequently approach houfes, and fettling on a rail, or poft, utter a piercing and difagreeable cry, which has been thought to refemble the words, *Whip-poor-Will*, the firft and laft fyllables pronounced the loudeft. This noife they repeat with little inter-miffion until the next morning, flying about from one place to another.

It is very difagreeable, and difturbing; and being fometimes repeated by four or five affem-bled together, and reverberated and multiplied by echoes from the mountains, it is difficult to fleep in their neighbourhood.

They are moftly on the mountains; they make no neft, but lay their eggs in an expofed fituation, upon the bare ground.

The GRAND GOATSUCKER;

Is the fize of a fmall buzzard, and inhabits Cayenne. Like others of the fame genus it is folitary, retiring into the hollow of a decayed tree in the day-time, and preferring one fituated in the neighbourhood of water.

K 5

Th.

The BLACK-BIRD; *an Elegy.*   By Mr. JAGO.

THE Sun had chac'd the mountain fnow,
   And kindly loos'd the frozen foil ;
The melting ftreams began to flow,
   And ploughmen urg'd their annual toil.

'Twas then, amid the vocal throng,
   Whom nature wakes to mirth and love,
A Black-bird rais'd his am'rous fong,
   And thus it echo'd thro' the grove.

O faireft of the feather'd train !
   For whom I fing, for whom I burn,
Attend with pity to my ftrain,
   And grant my love a kind return.

For fee the wintry ftorms are flown,
   And gentle Zephyrs fan the air ;
Let us the genial influence own ;
   Let us the vernal paftime fhare.

The Raven plumes his jetty wing,
   To pleafe his croaking paramour ;
The Larks refponfive ditties fing,
   And tell their paffion as they foar.

But

But truft me, love, the Raven's wing
   Is not to be compar'd with mine;
Nor can the Lark fo fweetly fing,
   As I, who ftrength with fweetnefs join.

O! let me all thy fteps attend!
   I'll point new treafures to thy fight;
Whether the grove thy wifh befriend,
   Or hedge-rows green, or meadows bright.

I'll fhew my love the cleareft rill,
   Whofe ftreams among the pebbles ftray;
Thefe will we fip, and fip our fill,
   Or on the flow'ry margin play.

I'll lead her to the thickeft brake,
   Impervious to the fchool-boy's eye;
For her the plaifter'd neft I'll make,
   And on her downy pinions lie.

When, prompted by a mother's care,
   Her warmth fhall form th' imprifon'd young;
The pleafing tafk I'll gladly fhare,
   Or cheer her labours with my fong.

To bring her food I'll range the fields,
   And cull the beft of every kind;
Whatever nature's bounty yields,
   And love's affiduous care can find.

And

And when my lovely mate would ſtray,
 To taſte the ſummer ſweets at large,
I'll wait at home the live-long day,
 And tend with care our little charge.

Then prove with me the ſweets of love,
 With me divide the cares of life;
No buſh ſhall boaſt in all the grove
 So fond a mate, ſo bleſt a wife.

He ceas'd his ſong. The melting dame
 With ſoft indulgence heard the ſtrain;
She felt, ſhe own'd a mutual flame,
 And haſted to relieve his pain.

He led her to the nuptial bower,
 And neſtled cloſely by her ſide;
The fondeſt bridegroom of that hour,
 And ſhe the moſt delighted bride.

Next morn he wak'd her with a ſong,
 " Behold !" he ſaid, " the new-born day !
The Lark his mattin peal has rung,
 Ariſe, my love, and come away."

Together through the fields they ſtray'd,
 And to the murm'ring riv'let's ſide,
Renew'd their vows, and hopp'd and play'd,
 With honeſt joy and decent pride.

When

When oh ! with grief the mufe relates
  The mournful fequel of my tale;
Sent by an order from the fates,
  A gunner met them in the vale.

Alarm'd, the lover cry'd, my dear,
  Hafte, hafte away, from danger fly;
Here, gunner, point thy thunder here;
  O fpare my love, and let me die.

At him the gunner took his aim ;
  His aim, alas ! was all too true :
O ! had he chofe fome other game !
  Or fhot, as he was wont to do !

Divided pair ! forgive the wrong,
  While I with tears your fate rehearfe;
I'll join the widow's plaintive fong,
  And fave the lover in my verfe.

The GOLD-FINCHES ; *an Elegy.* By Mr. JAGO.

'TWAS gentle fpring, when all the tuneful race,
  By nature taught, in nuptial leagues combine,
A Gold-finch joy'd to meet the fond embrace,
  And heart and fortune with her mate to join.

Through nature's fpacious walks at large they rang'd,
  No fettled haunts, no fix'd abode their aim ;
As chance or fancy led, their path they chang'd,
  Themfelves in every vary'd fcene, the fame.

All

All in a garden, on a currant-bush,
  With wond'rous art they built their waving feat;
In the next orchard liv'd a friendly Thrush.
  Not diftant far a Woodlark's foft retreat.

Here bleft with eafe, and in each other bleft,
  With early fongs they wak'd the fprightly groves,
'Till time matur'd their blifs, and crown'd their neft
  With infant pledges of their faithful loves.

And now what tranfport glow'd in either's eye!
  What equal fondnefs dealt the allotted food!
What joy each other's likenefs to defcry,
  And future fonnets in the chirping brood!

But ah! what earthly happinefs can laft?
  How does the faireft purpofe often fail?
A truant fchool-boy's wantonnefs could blaft
  Their rifing hopes, and leave them both to wail.

The moft ungentle of his tribe was he;
  No generous precept ever touch'd his heart:
With concords falfe and hideous profody,
  He fcrawl'd his tafk, and blunder'd o'er his part.

On barb'rous plunder bent, with favage eye,
  He mark'd where wrapt in down the younglings lay,
Then rufhing, feiz'd the wretched family,
  And bore them in his impious hands away.

                                          But

But how fhall I relate in numbers rude,
   The pangs for poor Chryfomitris decreed !
When from a neighb'ring fpray aghaft fhe view'd
   The favage ruffian's inaufpicious deed !

So, wrapt in grief, fome heart-ftruck matron ftands,
   While horrid flame furround her children's room !
On heav'n fhe calls, and wrings her trembling hands,
   Conftrain'd to fee, but not prevent their doom.

O grief of griefs ! with fhrieking voice, fhe cry'd,
   What fight is this that I have liv'd to fee ?
O ! that I had a maiden Gold-finch died,
   From love's falfe joys, and bitter forrows free !

Was it for this, alas ! with weary bill,
   Was it for this, I pois'd th' unwieldy ftraw ?
For this I pick'd the mofs from yonder hill ?
   Nor fhunn'd the pond'rous chaff along to draw ?

Was it for this, I cull'd the wool with care ;
   And ftrove with all my fkill our work to crown ?
For this, with pain I bent the ftubborn hair,
   And lin'd our cradle with the thiftle's down ?

Was it for this my freedom I refign'd ;
   And ceas'd to rove from beauteous plain to plain ?
For this I fat at home whole days confin'd,
   And bore the fcorching heat, and pealing rain ?

Was

Was it for this my watchful eyes grew dim?
  The crimson roses on my cheek turn pale?
Pale is my golden plumage, once so trim;
  And all my wonted spirits 'gin to fail.

O plund'rer vile; O more than Weezel fell!
  More treach'rous than the Cat with prudish face;
More fierce than Kites with whom the furies dwell,
  More pilf'ring than the Cuckow's prowling race.

For thee may plum or goosb'ry never grow,
  Nor juicy currant cool thy clammy throat;
But bloody birch-twigs work thee shameful woe,
  Nor ever Gold-finch cheer thee with her note!

Thus sang the mournful bird her piteous tale,
  The piteous tale her mournful mate return'd;
Then side by side they sought the distant vale,
  And there in silent sadness inly mourn'd.

DIRECTIONS for placing the PLATES in Part VI.

A CATALOGUE

www.ingramcontent.com/pod-product-compliance
Lightning Source LLC
Chambersburg PA
CBHW021510210326
41599CB00012B/1203